TEN YEAR STINT

Also by Lord Robens:

HUMAN ENGINEERING (Jonathan Cape)

LORD ROBENS

Ten Year Stint

THE PROFESSIONAL LIBRARY

First published 1972

This edition published by
THE PROFESSIONAL LIBRARY LTD., 2 SAVOY HILL, LONDON WC2
by arrangement with Cassell & Company Ltd.

Printed in Great Britain by
The Camelot Press Ltd, London and Southampton
F. 1171

Contents

1

From the House to Hobart House

I got off the night sleeper at King's Cross, walked over to the newspaper stand and bought my copy of the *Guardian*. There was nothing unusual in that. I had been doing it on and off for fifteen years—in fact, ever since I entered the House of Commons for the Wansbeck Division of Northumberland in 1945. What was unusual was the shock that was waiting for me. I could hardly believe my eyes. So this closely guarded secret known only to a bare handful of people was out.

It was there all right, no argument about it. The date was 13 June 1960.

The headlines on the main story on the front page were: 'N.C.B. CHAIRMANSHIP FOR MR ROBENS—Effort to take coal out of party controversy—LABOUR LOSING A "STAR" '.

The first paragraph of the story by John Cole, then the paper's Labour Correspondent and now Assistant Editor, read: 'An announcement that Mr Alfred Robens, M.P., will be the next chairman of the National Coal Board is expected to be made within a few days. Since Mr Robens is a senior member of the Labour "Shadow Cabinet", and a possible successor to Mr Gaitskell if he should have to resign the leadership of the party, the appointment will have political consequences.'

Cole, who was later awarded the Granada Press Scoop of the Year Award for his story, went on: 'The appointment is surprising for two reasons. One is that a Conservative Government should select the Labour Party's chief spokesman on industrial affairs and a strenuous political opponent to run one of the most important nationalised industries. This may be seen as an effort to take coal-mining, which has a difficult future even without any distractions, out of the world of party controversy. But it will scarcely be taken kindly by the "fuel fury" group on the Tory back benches.

'Even more comment will be caused, however, by Mr Robens's decision to go out of politics at a time when, inside the Labour movement, his star is definitely in the ascendant. The choice must clearly have been a difficult one. Mr Robens has a deep interest

I

both in the coal industry and in industrial relations. He was Parliamentary Secretary to the Ministry of Fuel and Power from 1947 to 1951—and thus saw the first experiment in nationalisation —and he was Minister of Labour during 1951.'

Cole was right about my reasons for being attracted to the job. Indeed these were my abiding political interests—mining and men at work.

But that did not prevent the storm breaking around my head. I had hardly a friend. I got it from all sides. There were swift comments from the unions, from the Press and from the politicians.

Within twenty-four hours of the disclosure in the *Guardian*, Will Paynter, the Secretary of the National Union of Mineworkers, told the Press: 'If Robens is appointed to carry out the Government policy of decentralisation, this will be met by most bitter opposition from this Union.' The South Wales miners also reacted promptly and so did Abe Moffat, President of the Scottish miners. They, too, were undoubtedly anti-Robens.

The President of the British Association of Colliery Management, Jim Bullock, quickly protested about my appointment to the Minister of Power and wrote a letter to *The Times* in which he said: 'Management's criticism of the appointment is the appalling idea that in this industry there is no person capable of handling the business of this great, complex and complicated industry; but that someone from outside the industry with nowhere near the experience can do it better than those with a lifetime's experience.

'My union represents men of the highest skills and experience in the industry, men of high integrity and calibre, and this appointment comes as a distinct shock to all of them and makes nonsense of my union's efforts to establish an appointments procedure in the industry which would conform to the high principles for which in the past this country has so rightfully won great respect.

'It is in this context that the appointment of Mr Robens appears to us to be wrong.'

Bullock had sent a copy of this letter to the Minister of Power. Less than a year later he wrote again to the Minister saying that, although the reasons for his organisation's objection and the principle behind it were unchanged, he had now had a chance to work with me, and that I had removed all the doubts in his mind

2

about my ability to do the job. He had, he said, been particularly impressed by my patience and understanding.

I very much respected Jim Bullock for that. It was quite typical of his blunt Yorkshire forthrightness, and my respect for him never waned.

Two of the three unions associated with the mining industry had commented: the third was yet to come and it was a pleasant surprise. The last important union, the National Association of Colliery Overmen, Deputies and Shotfirers, waited just over a week before they made their views known. Laurie Wormald, then President of the Association, said that I would have the full support of his members. They, too, would resist any suggestion of decentralisation. They also made it clear they would have welcomed an appointment from inside the industry. Having said that, though, he went on: 'We have the satisfaction of knowing that Mr Robens is a supporter of nationalisation and in these circumstances we feel that we can accept the appointment with respect and understanding. Knowing Mr Robens' background we are sure he will do his best to fulfil his obligations to the Government and the people in the industry.' I liked the spirit of that comment, of course.

The newspapers were able to find other reasons for criticising the appointment. The *Guardian*, for example, thought this was an astute move on the part of the Prime Minister to weaken the Labour Party in Parliament.

I am sure that this was not the case, but years later in his memoirs (published in 1971) George Brown, after describing me as 'an outstanding chairman of the National Coal Board', goes on to say that he always thought it a tragedy that I left party political life when I did. George continued: 'His loss to the Party was, I suppose, another part of the price that we paid for the Bevanite squabbling that lost us the General Election in 1959. . . . Had Robens stayed in the Parliamentary Party there is little chance that Harold Wilson would ever have been Prime Minister.'*

Nye Bevan died just before I left the Commons and Gaitskell soon after. It is hypothetical, but George would have been proved right if I had stayed in party politics. I would have had the support of most trade union MPs, of the 'sensible centre' as

* Lord George-Brown, *In My Way*. London: Victor Gollancz, 1971.

3

Ray Gunter used to call it, and perhaps of some of Gaitskell's Hampstead set.

Like most other politicians, I yearned to be Prime Minister. But I have never regretted my decision to go to the Coal Board. It was the right choice in the circumstances of the time.

The *Daily Telegraph* said flatly at the time the appointment was announced that it was not a good one, though they did add that they had nothing against me personally. Their hostility was based on the fact that the Committee under Dr (later Lord) Fleck, which had reported on the Coal Board's organisation a few years earlier, had said that the full-time members of the Board must normally come from within the industry. Furthermore they thought this principle should apply especially to the Chairman of the Board.

Though several papers chided the miners' leaders for objecting to the appointment, no newspaper was unreservedly in favour except the *Financial Times*, which said the appointment was a good and imaginative one. Curiously enough, the next warmest welcome was from *Tribune*, which may have surprised a lot of its left-wing readers by refusing to join in the criticism. 'If Alfred Robens shows that he is fighting to give the industry the fair deal it deserves, he should be able to command great support from the miners and the public,' was *Tribune*'s opinion. There was a similar surprising alliance between the Left and the Right at Westminster. Several miners' MPs put down Parliamentary questions and Richard Wood, the Minister of Power, faced a very angry attack from back-benchers of his own Party when he addressed a meeting of the Tory Power Committee. Their argument was that I had had no practical experience of controlling a large-scale industry, and that a man so closely identified with Labour policies should never have been chosen.

Hugh Gaitskell let it be known that he was in favour of my accepting the job and tried to dispel fears that this was to be the beginning of a Tory plot to wreck the Labour Party in the House of Commons by seducing some of his most prominent colleagues. It was characteristically generous of Gaitskell to try to be helpful in this way. After his statement, several other Labour MPs came out in support of the appointment and took the view that it was sensible to have state industries run by people who, like me, believed in nationalisation. Emanuel Shinwell, a former

Minister of Fuel and Power, with his usual keen appreciation of his Labour colleagues, thought it was a bad appointment.

The National Union of Mineworkers had for some time been anxious about what seemed to be authoritative hints that the Government were considering decentralisation of the industry. In the next few weeks there was a good deal of talk about a 'secret plan' to do this, and it was the main subject of anger at their annual conference. The Union had always been violently opposed to setting one coalfield off against another. Indeed, the need to prevent competition from the more profitable coalfields from putting men out of work in those that were more expensive to work, had dominated the miners' case for nationalisation throughout the 1930s.

As soon as my appointment was officially announced, I held a press conference at the House of Commons, where I gave a firm promise that there would be no interference with the existing system of national wage negotiations. I made it clear that I understood the miners' fears of any return to the pre-war system of district prices and even district wages. I said that everybody realised that there could not possibly be any decentralisation on broad policies like wages.

But I made it clear that I had a completely open mind on the decentralisation of organisation. Indeed, I defined my policy as being based on the principle that only those matters that could not be dealt with in the coalfields should continue to be handled at the centre. Any changes that were made would be for the sole purpose of improving the efficiency and profitability of the industry.

I was certainly not without a good deal of knowledge of the operations of the National Coal Board and the unwieldly nature of its administrative structure. In its first formative years I had, as Hugh Gaitskell's Parliamentary Secretary for four years from 1947 to 1951, been closely involved in its affairs. After that, in Opposition, I had kept a very close eye on the operations of the Board and could see that the whole situation of coal had changed in the energy market. But, instead of meeting the change, which had long been predicted, with new and dynamic policies, the Board appeared to have become demoralised at the top and this malaise quickly spread throughout most of the coalfields. In the midst of this change, due to fierce competition from oil, the untruth was circulated that the Tory Government was pressing the Coal Board to return to coalfield competition, with all that that involved. Certainly as far as I was concerned nothing could have been further from the truth. At no time before my appointment, when discussions were taking place, or in the period that followed under a Conservative administration, was it ever suggested to me that the industry should be reorganised in such a way. The Government were equally clear about my own policies, which could be expressed quite simply, of having no decisions made at the centre that could with advantage be made down the line. Looking at it from the outside it had always appeared to me that the Board's organisation was too highly centralised.

Until however I actually occupied the chair, I would not be able to assess to what degree decentralisation could be effected. Meanwhile, I took advantage of every opportunity I could to destroy this myth about the changing of the wage-negotiating machinery.

In a television broadcast I made it clear that I was a believer in decentralisation, but again emphasised that nothing would be done to break the system of national wage agreements. Will Paynter clamoured for the Government to 'come clean about their intentions' and claimed to have evidence that the Government scheme included the transfer to Divisional Coal Boards of responsibility for trading; the creation of a new federal type of National Board, consisting of Divisional Chairmen; stricter Parliamentary control of the industry's capital; and threatened investigation of those nationalised industries that failed to reach their financial targets. Quite clearly, someone had deliberately sought to stir up the Union with hair-raising stories about what was going to happen. The whole issue caused needless anxiety and was a threat to the confidence of the industry. Whoever it was who started this off was no true friend of the industry and certainly did nothing to help me have a smooth start in those important first few weeks. Which may well have been the reason for the campaign. When the knives are out in the Labour movement, they are used with some force, and that is when you need friends to protect your back from your supporters of yesterday.

The row over decentralisation continued to rage up to the time of the 1960 annual conference of the National Union of Mineworkers held, as always, during the first week in July. Several newspapers claimed to have seen copies of the secret plan, although there were many different versions of what it was supposed to contain. However, the storm around my head gradually blew itself out. I could not say, though, I was entirely unprepared for the row that followed the announcement, because, on the Saturday before the *Guardian* announced my appointment, I had attended the Northumberland Miners' Gala. As I sat on the platform as a local MP, I listened to speech after speech from the miners' representatives condemning the chap who was going to be the next Chairman of the Coal Board. There was no doubt that there was great hostility in the air and even at that time I thought there had been a leak in an attempt to embarrass me for some reason. Who was responsible I could only guess, but I believe the leak took place at that Gala. Whether I am right or wrong in my surmise as to how John Cole came to scoop the pool only he knows and he, in accordance with journalistic principles, is hardly likely to tell.

Whatever may be the facts about the leak, my appointment apparently provoked all sorts of sinister thoughts in people's minds and everybody seemed very eager to jump to conclusions about the implications of it.

But why Harold Macmillan invited me to the chair of the National Coal Board, to this day I haven't the least idea. I've never asked him; for me it has always been sufficient that he did so and I accepted.

At the conclusion of a meeting in the House in a committee room upstairs, one evening in May 1960, I left in company with colleagues to find Rab Butler and Dick Wood awaiting me. Butler asked me to come to his room, and after a few pleasantries Wood left Butler and me to talk there on our own.

Rab discussed the coal industry and its problems and the forthcoming retirement of the then Chairman and Deputy Chairman. I appreciated the problem of losing both at the same time, and in answer to his questions I suggested the names of about half a dozen people whom I thought would meet the requirements. I also argued that substantial changes were required in the administrative structure and launched into a detailed description of how I viewed things in the light of the very much changed energy situation which obviously was going to mean a smaller industry than planners had proposed and for which the Coal Board had made substantial capital investment.

We discussed names again and to my surprise he quietly said: 'The Prime Minister has asked me to discover whether, if you were offered the Chairmanship of the Coal Board, you would accept it.' I was quite taken aback. Leaving the political arena was farthest from my mind, but to be able to have a go at making the first nationalised industry a success from every angle—social, economic and political—that certainly was a challenge.

My reply was that I would not turn it down out of hand. I would consider. Meanwhile I was bound to secrecy.

When I saw the Prime Minister I had made up my mind to accept, but I said that before doing so, I should want to consult Hugh Gaitskell as Leader of the Labour Party. The Prime Minister's view was that this was reasonable, but before I did so he would wish to see Hugh Gaitskell himself or write him a letter. Meanwhile the pledge of secrecy was to be kept.

The letter was never sent. John Cole's scoop and disclosure in the *Guardian* had overtaken us, and here was I on a June Monday morning, facing a quietly angry Gaitskell who had learned from the *Guardian* this startling news about someone who had been more closely associated with him than any other man in the Party. You can't work with a man as I did with Gaitskell at the Ministry of Fuel and Power for four years, without building a close bond of friendship, bound up in an identity of political interests. We lived near to one another in Hampstead and saw each other frequently away from the House. He was hurt—perhaps more hurt than angry—and said that if I had approached him earlier he would have counselled my staying in the House of Commons. He saw the problem, upon my explanation of the pledge of secrecy, my agreement not to discuss until the Prime Minister had either spoken or written to him, and how the whole thing had gone awry by the *Guardian* disclosure. He understood, as one would expect from Hugh Gaitskell, but at the same time he made me promise that, if Labour won at the next election, I would be ready to join his administration. I agreed to do so if at that time he still felt the same. Sadly, by the time the next election came around, he was no longer with us. A cruel fate robbed the Labour Party and the country of an outstanding man who had the rare combination in a political leader of courage and complete integrity. In the event I served a second term at the Coal Board.

I said goodbye to the House of Commons at the end of the

session before the long recess and joined the Coal Board as Deputy Chairman and Chairman Designate on 1 October 1960.

Ten Ministers, four Permanent Secretaries, and ten years and nine months later, I walked out of Hobart House for the last time. The end of the stint.

For my first four months with the Coal Board my predecessor Jim Bowman and I worked in double harness. For me it was a time for getting to know the industry more intimately, assessing its potential, and working out the priority policies.

When I spoke to the Press on my first day at Hobart House, I made it clear that I was very conscious that, including the miners, their families, people who supplied the industry and people in the service industries in mining areas, there were altogether something like 3,500,000 people dependent for their economic livelihood on the size and prosperity of the mining industry.

The most urgent objective was for the industry to be given a market target to aim at. I had no hesitation in saying that this should be about 200 million tons of coal a year—a figure I had used in speeches before taking up my appointment. A figure to which, incidentally, the Labour Party was committed.

Two weeks after taking up my appointment I left for the first visit on a tour which would cover every coalfield. I spent two days in each Division, talking to Divisional Chairmen, their Boards, and senior officials down to Area level. This was to give me a chance to assess the quality of the people occupying these posts, to get their views on the prospects of the industry, and to establish myself as the future Chairman.

My tour of the coalfields and discussions with experts both inside the industry and among our customers and suppliers had, long before I finally took over the Chairmanship, convinced me that what the Coal Board needed was a massive sales campaign, backed up by the greatest mechanisation drive the industry had ever seen and a complete administrative reorganisation. The need for the sales drive was obvious. Coal had been losing business for several years, stocks had been piling up at the pithead, and competition from fuel oil was intense.

On the production side, I could see that new machines capable of carrying the industry to unheard-of heights of productivity were already available. What was needed was a determined and sustained effort to bring them rapidly into the pits and with goodwill on the part of the men. I could see that this was something the unions would support wholeheartedly and, to this day, there has been no Ludditism in the industry. I therefore invited Sid Ford, President of the NUM, and leading officials of the other miners' organisations to join me when I launched the sales campaign early in April 1961. In my speech at that conference, I pointed out that the industry placed orders with British manufacturers for machinery and supplies to the value of £250 m. a year, and put £500 m. into the pockets of British workers. Very little of that, I said, was spent in the Persian Gulf. In the industrial field, coal had suffered from the fact that it was less fashionable than oil. In many cases, even where local authorities were involved, coal was not being given a chance to quote for business. The attitude of some councils I found infuriating. On the one hand they went cheerfully ahead putting in oil to replace coal under the boilers of swimming baths, schools, town halls and so on. And then in the next breath, authorities in mining areas, who collected large sums of money in rates from the coal industry, were moaning because we closed pits.

I made it clear that the benefits that would flow from greater mechanisation must be shared by the consumers and by the miners. Mechanisation in 1960 accounted for 38 per cent of the total output, compared with just under 32 per cent the year before, though this generalisation concealed the abysmal lack of mechanisation in all Divisions except East Midlands, where, in producing 20 per cent of the deep-mined output of the country as a whole, they had reached 62 per cent power loaded. A wide range of machines suitable for most British conditions had been developed by the Board and by the manufacturers. What was now needed was a massive push to get those machines widely introduced.

I did not seek to 'knock' the market competition but rather to sell coal on its own positive advantages. To the domestic consumer we were able to promise a wide range of new and better appliances, improved distribution, and greatly increased expenditure in coal preparation plants.

During my coalfield visits as Chairman Designate I discussed my ideas for the industry's objectives and tested them on people. First, I usually talked about the marketing problems, leading on to the 200 m. tons aim. But I emphasised that we should have to fight hard for this share of the market. I went to some lengths to encourage the senior management to have confidence in our combined ability to run the industry ourselves and not as an extension of a Government department. We were not to be hand-maidens of the Civil Service; we would act commercially and not tolerate any interference from the Government. If the Minister wanted to give the Board a directive, then it would have to be a statutory one under the Act, for publication in the Annual Report, so that the distinction between purely commercial decisions and political decisions could be clearly seen. We never did have a directive of any kind.

Judging from the expressions on their faces and from their statements, the senior people in the industry liked what they had heard.

Then I came to my main point—greater authority for the management levels below Headquarters. Naturally, everybody looked pleased about this prospect, but then I went on to inform them that they would find in me the toughest taskmaster they had ever had to deal with. Personal responsibility was something from which there was no escape.

These first attempts to re-define the industry's purposes went down well. People obviously enjoyed being able to discuss the big policy questions, and for my part I was assessing the quality of those in the top jobs.

It was very evident, too, from the enthusiastic manner in which senior management in the coalfields joined in the discussions, that communications between the Board and the coalfields were not particularly good, and that management relished the opportunity of presenting their views based upon their day-to-day experience. This was intensely valuable to me because I was able on this first time round to get a pretty good thumbnail sketch of each of the forty-five Areas. It made me realise, too, that this tour could not be regarded as a once-and-for-all job kept for special occasions. It would be essential to continue quite regularly not only to visit Areas but also to extend the visits to collieries, so that I could not only see what was going on underground and meet the men on the job, but also pore over the pit plans with the colliery manager to see what the planned developments looked like. The colliery visits increased considerably in the years that followed, because it was here that the major part of the capital investment had to be made, and it was on successful colliery management that the future of the industry depended. The only thing that rang the cash register in the Coal Board was the coal coming up the shaft.

It became rapidly clear to me that we faced a very formidable task. Stocks of coal on the ground were huge and they had quite suddenly appeared. The peak was reached in November 1959, when there were 36·2 million tons of undistributed stocks. One miners' leader referred to them as the 'Mountains of Mourne', and they had a disturbing effect on morale. They were down to 35 million tons by the time I reached the Board but still represented a great dilemma. Markets were rushing away on the advancing tide of cheap oil and we were over-producing. If prices were to be got down or even kept stable, then mechanisation had to go ahead with all speed. Failure to do so would make coal even more uncompetitive. But mechanisation for a lower production would need less men at a time when pits were being closed because of over-production.

The nettle had to be seized—if we didn't mechanise quickly, even more pits would have to be closed. There was only one course to pursue—mechanise as hard as possible and keep a strict control over recruitment so as to enable natural wastage to take the load. A difficult decision, because this could not be a permanent policy. If we did not recruit the youngsters, the industry would bleed to death.

At the same time, there would be the biggest campaign to market coal that had ever been launched.

At the end of 1960 there were 698 pits; ten years later there were 292. When my term of office started, there were 583,000 people on the colliery books; when it ended, only 283,000. Undistributed stocks of coal were 35 m. tons at the start and 7 m. (mostly suitable only for power stations) at the end.

That this fantastic reduction in manpower and closure of 406 pits was achieved without losing a ton of coal in industrial disputes for that reason (we lost coal from other causes) is a tribute to the unions and management in the coalfields throughout the United Kingdom. I doubt if any industry anywhere in the world could point to such a success in industrial relations. Without the understanding and co-operation of the unions and of the men themselves, this task could never have been accomplished. I shudder to think what the consequences would have been for the nation and the industry had it been otherwise. A coal industry locked in industrial struggle would have brought the nation to its knees. Even today coal still provides 47 per cent of the nation's energy.

When I assumed the Chair on 1 February 1961 I had had four months of personal study in depth of the industry and I could quite clearly see the problems facing us.

As I sat in my chair in my office making notes of things to do, and things to think about, I could not but reflect how much easier it was to be a politician telling people what they should do and how much more difficult it was to do it.

With this grim reflection, my ten year stint began.

2

Countering the Militants

I suppose I must have travelled, as we call an underground visit, about three hundred and fifty pits in the course of my Chairmanship and met and eaten with thousands of men in the process. It was my invariable custom to follow an underground visit by a buffet lunch, if it was a morning shift, or a tea if I was there in the afternoon.

Invited were all the members of the pit consultative committee, the officials of the local branches of the National Union of Mineworkers and the National Association of Colliery Overmen, Deputies and Shotfirers (supervisors, with special responsibilities for safety), and the management, usually a group of about thirty or forty people altogether. I never got much to eat myself, as I made sure that I talked individually or in a small group to everyone in the room. Before I finally disappeared and the food was about over, I used to say a collective goodbye in a ten-minute speech about the latest happenings in the industry, adding a quick résumé of general policy and finally a rundown on the pit and its prospects.

From these visits I learned a great deal about the things that were uppermost in the minds of the men and management, and they were able to ask me all the questions that were troubling them. This was of inestimable value; all sorts of minor problems that, if left unanswered, could well have developed into big ones were resolved. I was very much struck by the lack of adequate communication between the Union rank-and-file membership at the pit and their Area and Headquarters officials. Certainly problems that were under active discussion at Area and National level were very often raised with me, which showed plainly that this communication gap existed. Subsequently the NUM executive produced a monthly newspaper called *The Miner*, which did something to help. It wasn't very attractive, but it was a considerable improvement on local papers like the *Scottish Miner*, which was plainly a propaganda sheet and unread by most Scottish mineworkers.

This close association with the men underground and on the surface at the buffets gave me a remarkable insight into the character and quality of the miners. They are a grand crowd—what my grandmother would have called 'gradely folk'. When subsequently I had my violent disagreement with former Labour colleagues on pit closures, and argued about the social and economic impact upon the miners and their families, I was totally sincere—I meant it. It was not mere lip service on my part when I paid tribute to the quality and character of the mineworkers. I really was proud of them and I always felt privileged to lead an industry which contained such fine people. My personal relationship with the men and their leaders was schizophrenic—a sort of love–hate relationship.

Face to face—no suppressed grumbles, no unresolved problems, honest understandings, even if on some occasions, differences. The Communists and the militants, of course, hated me. I was a thorn in their side, very largely because I had forgotten more than they ever knew about organisation and meeting tactics, and particularly the use of abrogated authority by which a militant handful could determine the decisions of large numbers. Whether they realised it, I don't know, but I had been active in the Trade Union movement since I was seventeen years of age and from 1927 onwards I had continuously in the Labour, Trade Union and Co-operative movements fought the Communist and the fellow traveller. (Actually, the fellow travellers were much more dangerous than the Communists: you knew where the Communist stood, but the fellow traveller was a wolf in sheep's clothing, never to be trusted.) I had opposed them right up to 1945, politically and industrially, and have done so ever since. But the Communists and their cohorts were a minority, albeit a minority that wielded a much greater influence than their numbers warranted, relying very largely on the deep loyalty of the mass of the men to the Union and the poorly attended Union meetings.

So the love–hate relationship manifested itself in some interesting ways. Down the pit or at a social gathering I never had anything but politeness and courteous and warm friendship from all whom I met. When I was in trouble with the Minister of Power and my removal from office (or resignation) was imminent, they rallied round and closed their ranks about me.

But on the weekend platforms, I came in for plenty of stick on all sorts of accounts. Pit closures, house rents, wages claims. I was the arch villain. In the end, when the time came for me to go, only the Communists and the militants were pleased and jubilant. The letters of regret from mineworkers came pouring in and with many of them came little tokens to refresh the memory.

Perhaps I was too starry-eyed in my youth, but I really did believe that, when public ownership replaced the old private coal owner, strikes would be a thing of the past and differences of opinion would be settled in a civilised manner around a table between men and management at every level. I could not understand why the men did not realise that the only effect of a strike was to lose them wages. No one in management ever lost a penny as the result of a strike. The loss of coal and the subsequent loss of income to the Board meant increased prices in due course. Increased prices lost markets, which in turn closed pits, putting men out of mining employment for good.

I have known many pits that were shut by reason of their being grossly uneconomic because of incessant industrial disputation and general bad behaviour on the part of Union lodge officials, which influenced the men. To me it was a crime and I was resolved as far as I was able to remove the cause of possible dispute.

It did not take long to discover that the system of piece-rates was the principal source of industrial strife. It only needed, in some cases, a dispute upon a rate affecting just a handful of men, and the whole pit employing some 1,500 to 2,000 men stood idle, most of the men brought out on strike not having the faintest idea why in fact they were out except 'it's summat to do with Seven's face'.

I have always been against the use of piece-rates as a means of wage payments in industry. I regarded the whole system as degrading. Perhaps it was because so much of my early life was spent in an environment of two to three million people unemployed. Twenty years of that and I saw at first hand the indecencies of the carrot and the stick. If the task finally fixed could be accomplished—and more beside—then the worker could have the carrot of higher earnings; if the older or less dextrous could not keep up, then there was unemployment and the pitiable dole to exist upon. And this is no exaggeration, this is the world in which I lived and grew up. As a full-time trade union official my life was bound up with the unlucky ones, not the lucky ones. Hours and hours at

Labour Exchanges pleading the cases of members before the Courts of Referees. Piece-rates are not something I can easily live with. It was fortunate for me, too, that Miners' Union policy was to achieve a day wage.

There were other reasons for getting away from crude methods of incentive payment. We needed to do two things: first of all, to predict our wage costs (which were well over half of total costs) with more accuracy than we had done in the past; second, with mechanisation, to secure an even greater degree of labour mobility from job to job within the pit and from hopelessly unprofitable to potentially profitable pits, sometimes in other coalfields.

The first task was indeed daunting. Traditionally, wages in the mining industry had been determined by piece-rates at the coal-face and by a variety of other practices for men employed else-where underground and on the colliery surface. I inherited a first-class job of work which had been done in tidying up, on a day wage basis, wages elsewhere underground and on the surface. But the real problem lay in regulating the piece-work wages of the 'face-workers', the aristocrats of the industry and those who turned up to Union meetings at pits. These were the wages that were most subject to local pressure and thus most defied national prediction. Equally, because these were negotiated locally, the power that went with the negotiations accrued to local Union officials, rather than to the National Union. Both the Union and the Board (not to mention the vast majority of Union members who were also the Board's employees) had therefore an interest in securing some form of rationale in the industry's wage structure. But it wasn't easy. The militant Left saw quite clearly that a new national wage structure would weaken their ability to create strife at pits. And they lived on strife. The real interests of the ordinary good rank-and-file miners were subordinated to the personal ambitions of the power-hungry Left.

The drive towards a new wages structure was given great impetus in 1964 by Will Paynter. He was actively opposed in this by the Communist Party and attacked (though not by name because he was just about the only trade union leader of distinction still left in the Communist Party) by the *Daily Worker*. His critics were still angry with him for recommending acceptance of a recent wages offer. Once again, they were demonstrating that they were the enemies of innovation and progress.

His authority was augmented rather than diminished by these attacks and Paynter addressed the NUM's annual conference about the need for a review of the wages structure. He argued that they should be concerned about the impact of the technical revolution then going on in the industry on the relationship between the different wage rates. Such developments would completely outmode the pay structure and cause great changes in the skill, responsibility and labour content of different jobs.

It was a short, sharp debate and the Union's Executive emerged with a completely free hand to carry out a fundamental review of the whole structure. Once again, Paynter's ability to give a real lead had worked for the good of his members and of the industry as a whole.

Paynter had spoken on the Tuesday. I addressed the conference on Thursday morning, as usual. I followed up Paynter's arguments about the need for a fresh look at wages. Pointing out that the gap between day wages and piece-rates was widening rapidly, I said we were producing an East End and a West End in the colliery villages. I promised that as soon as the Union were ready, we would sit down with them to solve the problem.

Bill Webber, the Board Member for Industrial Relations, and his team had the tremendous and long-drawn-out task of negotiating the details; the National Power Loading Agreement covering nearly 200,000 men was finally agreed in March 1966 with the Union's National Executive, who recommended it to their delegate conference the following month. This objective had been contained in the Miners' Charter drawn up just after the Second World War. Although the agreement did not bring parity immediately between the better-paid and the less well-paid coalfields, it provided that this should be achieved not later than five years from the date of the agreement.

If we had raised the rates everywhere to the level of the best, many pits in the less well-paid coalfields would have been crippled financially. It was necessary to persuade the best-paid men to mark time and even, in some cases, to accept cuts in wages. I do not believe any other union in the country has ever gone to its members with such a proposition. But the NUM, and in particular Will Paynter, did do this. It was certainly not easy for them to persuade some of their members to give up maybe £5 or even £7 a week so that their less fortunate colleagues in the unprofitable coalfields could have an increase. But this is what they did and it demonstrated once again the tremendous loyalty to Union decisions. Time and time again this loyalty was exhibited, but very often it was exploited by the militants who much preferred to break procedures and agreements than to honour them.

The immediate result of the change was a substantial reduction in unofficial strikes. In the first year of the new arrangements the loss of coal output through disputes fell from an average of nearly 1½ million tons a year to 439,000 tons (about 70 per cent), and in the following year (1968/69) to 328,000.

But comparative peace in the coalfields was not to the liking of the militants, so new sources for unofficial strikes had to be found; towards the end of 1969 the industrial relations situation boiled up again, and not unexpectedly in Yorkshire.

It is not a coincidence that today the Yorkshire strike record is such a bad one. For several generations, long before nationalisation, the most turbulent areas were Wales and Scotland; these were the Areas where production costs were high, largely because of low productivity. When the demand for coal took its disastrous dive in 1957, it was the high-cost pits that were closed first, and this meant substantial pit closures in both coalfields. Despite their militancy, or perhaps because of it, their wage rates were well below the Midlands, where comparative industrial peace prevailed and with it higher productivity and higher wages. In fact it was only the introduction of national day rates and the abolition of piece-rates that dragged the wages of the miners in Wales and Scotland from the lower level produced by militancy up to the higher standards of the Midlands which had been obtained by intelligent co-operation with management.

However, be that as it may, with the closure programme having a big impact in Scotland and Wales the militants could see that their influence in the Union would wane, and so a determined attempt was made to gain control of the Yorkshire coalfields. Activists moved into employment in the Yorkshire pits, and, over a comparatively short period of time, power at many pits was in the hands of men with anything but a Yorkshire accent.

The trouble-makers were not just vocally militant, but encouraged strong-arm methods and intimidation. I had my first taste of this within a month of my taking over the Chair. I had, in fact, to face two challenges to my authority. First, there was a widespread stoppage in Yorkshire over piece-work earnings. These were not negotiated nationally, and when I visited Leeds during the strike, I made it clear that I had no intention of intervening. Nor would I give way to force, anarchy and blackmail.

The real purposes behind the strike were not industrial at all. This was a blatant attempt by the Communists to win control of the Yorkshire coalfield which, in its turn, would have given them virtual control of the National Union, since they already dominated the Scottish and South Wales Areas. The Yorkshire strike was an ugly one. The Area Secretary, Fred Collindridge, spoke of acts of terrorism at some Yorkshire pits, and agitators toured the coalfield in flying-columns spreading the strike. Incidentally, this was a tactic that was to persist in disputes in this coalfield right up to the end of my Chairmanship. The stoppage was condemned by the elected Yorkshire Area Officials and the National Executive of the Union. But it lasted for two weeks and lost more than 800,000 tons of coal. However we could not give in to anarchy; we stood our ground and refused to negotiate until work had been resumed. The strike achieved nothing for the men who came out and they lost about £2 m. in wages. They also robbed the Board of about £4 m., which reduced our ability to be more generous at a later stage in wage negotiations.

Scarcely had the Yorkshire pits gone back to work than there was a threatened stoppage in South Wales. This time there was no attempt even to pretend that it was connected with the mining industry. The one-day stoppage, threatened for May Day, was planned as a protest against increased National Health charges announced by the Government.

Immediately the strike was suggested, the South Wales miners were warned that if they did stop on 1 May, they would lose two days' pay. The men had been accustomed to getting a bonus shift under an agreement if they worked five days in the week. However I made it clear that the new system, which had recently been introduced, giving the men one and one-fifth days' pay for each shift worked, would certainly not apply in the event of a strike for purely political purposes.

We used every means open to us to make this clear and to point out that a strike of this kind would only damage the industry and the men themselves. This time the rank and file rebelled against their militant leaders' intentions. The threat collapsed. The political strike had been abandoned, but it had been a near thing, and only averted by the commonsense of the mass of good, hard-working Welsh miners, who refused to be used as cat's-paws in the political game.

We were able to avoid major disputes for many years after that. In fact it was not until September 1969 that big trouble raised its head again in Yorkshire. Men at Cadeby Main Colliery, near Mexborough, were on an unofficial pay strike in defiance of Yorkshire leaders, and we were sticking firmly to our line of no talks with men who were on strike. An attempt to get nineteen collieries in South Yorkshire out in support of the Cadeby men failed; the other pits wanted them to go back to work to see if a satisfactory solution could be reached in negotiation. They were willing, if this failed, to consider afresh the possibility of supporting action.

While the Cadeby men were on strike, pickets from several coalfields demonstrated outside Hobart House in support of the claim for shorter working hours for surface workers. The pickets were accompanied by at least two members of their National Executive Committee and one Member of Parliament, Tom Swain of North-East Derbyshire. I have no doubt that the presence of these gentlemen gave much encouragement to the men who had dishonoured their bargains, broken all the negotiated procedures and defied their official leadership.

Nevertheless, negotiations can only sensibly be carried out with the official Union leaders, and our visitors were told that the Board could not discuss the issues involved with them. While the seven hundred or so demonstrators had taken the day off to come up to London I was out in the coalfields, and there was an attempt to argue afterwards that I had deliberately slighted them. In fact, I was carrying out an engagement made a long time earlier and I had no intention of breaking it to receive men who should have been at work themselves, leaving the negotiations to their officially elected representatives.

A few other pits did join the Cadeby men, but then this was overtaken by the main strike over surface hours which began in Yorkshire with 70,000 men out.

I recollect that, on the very day the Yorkshire pits stopped, a delegation from their Union went to the Ministry to try to persuade the Government to announce a start on building the second phase of the Drax coal-fired power station in the county. Hardly a propitious start! I warned that strike action could well jeopardise the industry's chance of getting this important decision. And in the same public statement—ironically enough issued from a weekend NUM school I was attending in North Wales—I said that whilst we would not parley with the leaders of unofficial strikers, no unofficial strike would deter us from continuing our negotiations with the official leaders of the Union.

There were the usual ugly picketing incidents. Miners at Water Haigh Colliery near Leeds defied the unofficial strike and went down the pit. They were given the fraternal ultimatum by the pickets that, if they didn't come up within an hour, there would be a strike of the winding enginemen who raised and lowered the cages in the shafts and this would mean that they would have to stay down the pit. Not surprisingly, they returned to the surface. Certainly, in an unofficial strike no man should be allowed to threaten others in this way.

Within a few days the strike had spread to some pits in Scotland, the Midlands and South Wales.

We were at this time negotiating with the NUM, not only over working hours, but also over wages. The Union had claimed an increase of 27s. 6d. a week per day-wage man and we agreed to meet this claim in full—every penny of it.

One would have thought that this unprecedented negotiating success by their leaders would have impressed the strikers. Not a bit of it. Leaders of the Yorkshire unofficial strikers immediately clamoured for the resignation of the two national officials of the Union, Sidney Ford and Lawrence Daly, the recently elected Secretary. In my opinion, the militant leaders could not be affected by our reasoned arguments.

At the height of the strike I took part in a discussion with miners from several coalfields on the BBC television programme *Panorama* with Robin Day in the chair. It was a pretty angry confrontation, and the miners, many of whom I had met at Union conferences and at their own pits, made great capital out of the fact that the recent strike of dustmen had obviously paid off. And there can be no doubt that the dustmen's strike, which brought them a substantial increase in pay, had a profound effect upon other workers, including miners.

No matter how skilfully we and the NUM leaders conducted our internal affairs, the industry could not be isolated from what was happening elsewhere. I know there was great bitterness in the Yorkshire pits when a new agreement covering bus staffs in the county gave women conductors a minimum wage of £18 a week—considerably more than our minimum for an adult male worker.

But it was still the most senseless strike of any size ever to affect the mining industry. We had agreed to give every penny of the increase the Union demanded. Maybe it was basically a protest against the way the industry had been treated. It was noticeable that few pits went on strike in coalfields where there were prospects of more closures to come. Yorkshire had very few collieries in jeopardy, and obviously some of the men there thought that the strike could not affect their own futures.

The Yorkshire flying-squad of pickets were up to their tricks again, but on at least two occasions they came unstuck. When they turned up at a colliery in Nottinghamshire they were greeted by about forty angry miners' wives. Apparently they couldn't face this and they left the colliery, Bilsthorpe, when they were greeted with cries from the women: 'Are you a worker or a shirker?' The cars quietly turned round and drove away.

A group of Yorkshire pickets that turned up at Sutton Manor Colliery near St Helens was confronted at the pit gate by the local NUM Branch Secretary, Joe Connolly, who has for many years had to use a wheelchair because of illness. Some of the Lancashire miners were extremely resentful of the activities of the pickets from over the county boundary, because the Yorkshiremen knew that their pits would reopen after the strike, but in Lancashire many collieries were in jeopardy because of their financial results. The men feared that their prospects would be still further threatened by the financial losses caused by the stoppage. Connolly told the Yorkshiremen this in no uncertain terms, and they turned round and drove away, presumably to try the same tactics at another pit.

Towards the end of the strike's second week, a movement back to work began. The militants knew they were beaten and asked the TUC General Secretary, Vic Feather, to intervene.

Feather was then trying to stave off the Labour Government's compulsory incomes policy by settling disputes himself, but I am afraid the only effect of his intervention was to save the militants' faces. The leaders of the unofficial dispute were able to claim that Mr Feather had produced a formula that allowed them to order the men back to work. He hadn't, of course. Always, militant leaders of unofficial strikes dread that the stoppage will end in such a way that their authority is affected and their future influence reduced.

As the strike ended, Britain's farmworkers made a complaint about miners on strike who took part-time jobs on the land. The fact is that ethics no longer mean anything in an unofficial strike. Men can apparently protest against their conditions and at the same time go off and take another job which saves them making any sacrifices for their principles.

The 1969 stoppage lasted a fortnight and lost us more than $2\frac{1}{2}$ million tons of coal. At its height 140 collieries out of the 307 then operating were involved, including all those in Yorkshire. The capture of the Yorkshire coalfield by the militant Left had been complete. Their victory had met with 100 per cent success. Not a pit in Yorkshire turned a wheel at the peak of the dispute. This time the financial cost to the Board was about £15 million and I had to ask the Government for permission to increase coal prices. With that and the wage increase another twist was given to the inflationary spiral.

When the rank and file were finally able to express their views on the Board's offer, the result was a massive majority in support of the National Executive Committee's recommendation that the offer should be accepted. In fact, nearly 194,000 voted for acceptance, and only 41,300 against. The question of surface hours, which was supposed to be the reason for the strike, was put off for further negotiations later on.

So militant leaders had been able to stop nearly half the pits in the country when this was clearly against the wishes of the vast majority of their members. Once again the loyalty of men to their Union as they saw it had been prostituted by the determined militants, who had in fact got nothing out of the strike and lost £3¾ million in wages as a consequence.

There was an unfortunate anti-climax for the NUM's Secretary, Lawrence Daly, ex-Communist, ex-Secretary of the Scottish area of the NUM, who is now one of the most articulate critics of the Communist Party, but certainly would never describe himself as a moderate. When he went to address a meeting in Yorkshire soon after the 1969 unofficial strike, he must have been taken aback by the violence of his reception. I do not mean that he was subjected to physical violence, but he was called a traitor and told to get out. All this because he had tried to argue that the Union's constitution must be preserved. Defending himself for not having supported the unofficial stoppage, he said that would have been a breach of the Union's rules. However he was reported to have said that the strike had had an impact and that it had taught the Coal Board and the Government that the miners' loyalty could not be taken for granted.

I could not admire his logic or his reasoning, though I understood the difficulty he was in quite well. He had been elected to national office on a militant programme, defeating the more moderate Joe Gormley, from Lancashire. Gormley probably suffered from being a member of the Labour Party's National Executive, thus being associated in the men's minds with the unpopular policies of the Labour Government towards the coal industry. Joe has since had his revenge by winning the more important job of National President, defeating a Communist from Scotland, Michael McGahey. Daly, having been elected on an aggressive policy, was caught between his responsibilities to the official leadership and his loyalty to his past supporters.

But this sweet victory for the Yorkshire militants was to turn to a bitter taste in the mouth as time went by. Emboldened by their success, they laid their plans for an even stronger attempt at the next round of wage negotiations. These disruptive elements, in my opinion the real enemies of the miners, could only succeed if they could prevent the democratic procedures of the Union from operating. In any ballot required under the Union rules they were completely defeated on every occasion. The rank and file always faithfully accepted the lead of their National Executive, which usually followed a sensible, moderate and constructive line.

The settlement of the wage claim in November 1969, which was the highest in the history of the industry, was soon overshadowed by settlements in other industries. Large inflationary wage settlements were the order of the day in other industries, and the feeling of being left behind could easily be created by continual reference to the declining position of the mining industry in the national wages' league.

In July 1970 the NUM Conference unanimously passed a resolution for a minimum weekly wage of £20 on the surface, £22 underground and £30 for men on the National Power Loading Agreement, and for good measure a resolution from South Wales for strike action until the Board conceded the claim was narrowly carried by 169 votes to 160. This meant increases of £5, £6 and from £3 to £7 on the NPLA rate; we offered £2 10s. 0d., £2 10s. 0d. and £1 17s. 6d. respectively at a joint meeting with the Union.

The South Wales miners' resolution was of course out of order, contrary to the rules of the Union and should never have been permitted. But the Chairman allowed the debate and Lawrence Daly carefully spelt out that any decision to strike could only be accepted within the rules of the Union and that meant a ballot vote of the membership and a two-thirds majority. The resolution was carried, though only by a majority of nine. But the narrowness of the majority was of no account; the strike resolution became a decision. The Communists and their militant supporters had won their first victory. They had virtually tied their negotiators down to £5 or else!

The Left were jubilant. Still smarting from the defeat by the commonsense of the rank-and-file miner during the previous October strike, they had moved a good deal nearer, by the Annual Meeting decision, to what they hoped would be an official strike.

On 16 September the Board and the Union met to resume discussion on the wage claim and the offer already mentioned was made. This was rejected as a 'disgusting offer'.

The militant Left made sure that their negotiators should be in no doubt about their views. The meeting was picketed by demonstrators mainly from South Wales, Scotland, Kent and Yorkshire. Estimated size of the crowd of demonstrators was eight hundred—all paid for out of local Union funds. South Wales produced the initiative for this, as they had provided the strike resolution. So far, so good, everything going to plan. But even greater success was to be theirs.

The lobbying coalfield demonstrators arranged a meeting at Conway Hall in London that evening. South Wales organised the meeting and invited the President and General Secretary to attend and report about the offer that had been made that day. Lawrence Daly and Sid Schofield both attended. I do not think that Daly minded in the least, and I have no doubt that Sid Schofield, who was Acting President in the absence through illness of Sidney Ford, attended this meeting with a very genuine desire to cool down the temperature that had been generated.

This meeting broke all tradition in the Union. It meant that a completely unofficial assembly received a report from its most senior officials before properly elected Executive Committee members could report back to the coalfields. Furthermore, the meeting took the decision out of the hands of the elected officials, once more leaving the National Executive Committee of the Union no room to negotiate or think again about the Board's offer. The meeting also provided the platform for co-ordinating the militant action that was to come.

When the National Executive Committee of the NUM met the next day they unanimously decided to recommend a national official strike. This decision, in accordance with the rules of the Union, had to be submitted to a ballot vote of the membership and obtain a two-thirds majority.

In the event the ballot paper was drawn up in such a way that the members were not allowed to vote on the acceptance of the wages offer of the Board but only upon the Executive's recommendation to strike. When I challenged Daly with the consequences he just laughed it off, but it was clear to me that he considered this a clever tactic, as he was, he said, in favour of strike action.

It was clear that the Union now was fairly in the grip of the militants, and the moderates had had to join the bandwaggon to preserve their position. Or so they thought. They had experienced the noisy demonstration of the trouble-makers but overlooked the calm, sensible thinking of the rank-and-file membership. But it was ever thus, and those who speak up against anarchy and the Communists are labelled as red-baiters; a few left-wing journalists in Fleet Street are good at this. The provincial journalists have always seemed to me much superior to most of their London colleagues in objective reporting.

It seemed to me that there was an obligation on the part of the Board to its employees to provide them with the facts of the situation. The rest must be a question of their own personal judgment. I was determined however that they should have the plain, unvarnished truth, and on 23 September a leaflet summarising the Board's offer and the reply made to the claim was sent by post to all mineworkers at their home addresses.

I received a fantastic number of letters from mineworkers and their wives thanking me for the information and complaining about the intimidation that was beginning to show its head in the coalfield. For every letter I received I knew a hundred more men felt the same.

There was however still a chance that commonsense could prevail, so on 28 September 1970 I wrote to Lawrence Daly, as General Secretary of the Union, reminding him that the NUM could still take the claim to the National Reference Tribunal, pointing out that the possibility of settling the question through the agreed machinery should be given careful consideration. But the South Wales miners had done their work for strife very well indeed. Daly's reply on 2 October was that the decision of the Annual Conference meant that there was an obligation to consult the membership and the ballot on the strike action would proceed. This was leadership from below with a vengeance. Daly was now the victim of his own advocacy and, although he did not know it, he was actually moving towards his own humiliation as I shall describe later in this chapter.

The letter to the mineworkers' homes was followed up by a special free issue of *Coal News* to all mineworkers. In it we were scrupulously fair to ensure that both the view of the Board and that of Daly were clearly and honestly expressed.

In contrast, a free copy of *The Miner* issued by the NUM was devoted entirely to calling for a strike, which was the recommendation of the National Executive Committee—it was in fact a blatant appeal to the traditional loyalty of the mineworker to support the National Executive Committee of the Union and ignored the real issue, which was whether the offer of the Board should be accepted or rejected.

The ballot took place from 12 to 16 October in a highly charged atmosphere in which Union officials in most Areas were now holding meetings urging the men to vote for a strike. Lawrence Daly stumped the country, stressing the importance of supporting the Executive by voting for a strike.

Yorkshire was the key coalfield. Here the maximum pressure was exerted to bring about a favourable strike vote. There was a full briefing of everyone in an elected position at each pit, leading to meetings for each pit in the local Welfare Hall, followed by a mass meeting for each of the Coal Board's Areas. This certainly had the look of careful psychological planning intended to whip feeling to a crescendo. Jock Kane, the Communist and Financial Secretary of the Yorkshire Miners, did most of the talking at the Area Meeting with Branch officials, and Lawrence Daly made the final impassioned plea at the winding-up meeting.

In Doncaster, where the last of the four meetings was held, there were all the indications of a determination to strike, whatever the result of the ballot.

The men were given advice on how to run unofficial stoppages, told that they must have an attainable objective and secure maximum improvements with minimum sacrifices.

The ballot itself was very far from secret. At all pits the management were instructed to offer the miners' lodges facilities that would enable a secret ballot to be held. Only in a few cases was this facility accepted. In Yorkshire at many pits the ballot took place on tables erected at the dirty end of the pit-head baths. Usually men coming out of the pit are anxious to get showered and changed, so there's little time to argue or reflect. At others the Committee arranged for the display of a huge enlargement of the voting paper showing where the cross should be placed, and Committee members leaned over the men's shoulders to see that they marked their papers in that way. And at one pit at least separate ballot boxes had been marked 'yes' and 'no'. This picture was similar throughout the coalfields, with some notable exceptions where the ballot was honourably and honestly conducted.

The results began to appear quite quickly, although the votes were to be sent to the Electoral Reform Society, who would count them and declare the result. However the papers had to be 'straightened' before being sent to the Electoral Reform Society, and an unofficial count frequently took place whilst this happened.

The general impression of the Yorkshire campaign was that it was unscrupulous and the ballot wide open to a number of manipulations from incorrect recording of the votes and intimidation at the point of ballot. The pattern varied in the coalfields as one would expect with the degrees of Communist and militant activity against moderation and reasonable judgments.

On 23 October the decision of the rank and file showed there was insufficient support for a strike. The results were announced:

FOR STRIKE ACTION	143,466
AGAINST STRIKE ACTION	115,052

This gave a majority of 55 per cent in favour of a strike with an 83 per cent poll.

The Executive had not received the two-thirds majority which was necessary for an official strike. Whilst South Wales and Scotland, both Communist-dominated, had secured a two-thirds majority in favour of a strike, Yorkshire had failed to do so, with 60 per cent in favour. In the English coalfields as a whole even a simple majority had not been obtained.

The figures, too, were revealing. The total vote with spoiled papers was just short of 260,000, and if this represented 83 per cent of the membership, as was reported by the Union, the membership at that time was 313,000. The ballot papers were issued with a 2 per cent margin for wastage and error, which means that 319,000 papers were in circulation.

For the week ending 17 October 1970, the number of men on colliery books (excluding under-officials who were not in the NUM) was 262,000; in addition there were 37,000 other mineworkers, including clerical staff members of the NUM. This allowed an actual total of about 300,000. Reconciliation of men employed with ballot papers issued may be complicated by the fact that in some Areas retired men were probably allowed to vote.

However, absence from work during the week of the ballot was high (perhaps because a lot of the militants were electioneering) at 21 per cent. This means that 60,000 men were not at work so that, unless all the rest rose from their sick beds, only about 240,000 could conceivably have voted. But strangely enough, according to the Union's figures, 260,000 men voted.

In Scotland absence was 22·5 per cent, so that 23,500 men at work produced 26,500 votes. In Wales absence was almost

24 per cent: thus, 31,000 men at work produced a vote of 35,400.

Over the coalfields as a whole the excess of votes over attendance was 8 per cent; in Communist-dominated Wales and Scotland it was 14 per cent and 13 per cent respectively. These are the facts; the reader must draw his own conclusions.

Rules and honourable bargaining are not in the book of the militant Left. Despite the adverse ballot, despite the rules, despite the fact that further talks were offered and accepted, unofficial strikes took place at many pits during the four-week period from 26 October to 23 November.

The renewed wage negotiation took place on Tuesday, 27 October. The meeting was protracted and lasted all day, finishing at 20.40 hours with a modest 10s. 0d. increase all round on the original offer. At midnight that night one face struck work at Brodsworth in Yorkshire, bringing the whole pit out the next morning. This was the pit where Curley Owen was the delegate, but it was Councillor Roy Harris, a chargeman in his unit, who led the men out. There was no meeting. The Doncaster Panel of the NUM met that evening, with four pits affected, and decided to strike from midnight.

From that moment complete confusion reigned. By that time all the other left-wing organisations, the International Socialists, the Socialist Labour League, the Institute for Workers' Control, together with the academic lefties from at least three Universities, and a technical college lecturer, plus an assortment of Marxists, Trotskyites, Maoists and others, non-miners in the main, joined in. The Communist Party lost control. They were themselves being attacked by the more extreme Left. They therefore found themselves advocating constitutional means of showing displeasure, that is, overtime bans, and finally strike action.

Pits were in and out until the third week, when in Wales the stoppage became total; all but eight pits in Scotland were idle; so were all the pits in Doncaster. The picketing became rough and intensive.

The Yorkshire Area Council meeting in Barnsley on 29 October was heavily picketed by men from Doncaster. The picketing failed in its objective and despite angry voices inside and outside the Council Chamber the meeting broke up in disorder with a resolute

President in Sam Bullough refusing to take a strike motion. Other signs of an ugly mood developed in Doncaster Area, where the threat was made to withdraw safety men—a precedent almost unheard of in the history of mining. At Glasshoughton Colliery a hundred pickets had to be removed from the premises by the police.

Those trying to get to work had the panels of their cars kicked in and crowds of pickets around the entrance to the colliery prevented men from going to work. The pickets were well organised, provided with transport and they used their own cars with free petrol. Most of them were paid. All stories of rough handling and intimidation were denied, of course, but when the pickets went further afield, the hooliganism could no longer be disguised.

Detailed reports reaching me from all over the country showed the extent to which an influential and determined body of men, regardless of the damage to the men they were purporting to help and the industry in which they worked, were prepared to go.

However, the end was in sight and the strike was melting when I went to Doncaster on Thursday, 19 November. The pickets concentrated on the NCB Area Headquarters to greet me as I arrived. All the principal characters were present, including Harris and Owen of Brodsworth. Owen was a particularly noisy character with whom I had had a brush on a previous occasion.

I left my car and went amongst them hoping to have a rational discussion. They were a 'yarling mob'—crude, vulgar and unfit to lead the decent men I know in the pits. How in heaven's name men like this can possibly be elected as leaders of good Yorkshire miners, I cannot understand. They objected to being described by me as a 'yarling mob', but that description was an understatement. But for the presence of the police I believe they would cheerfully have murdered me.

But this was the end. As one commentator said: 'It was perhaps appropriate that they should spend their last energies in this way —they started out spurred on by the inflammatory speeches of the National Leader and ended up empty-handed with the man who warned them that it would be so.'

When I spoke out on television and radio about these people

and accused them of being subversive and out to damage not only the industry, but the country as a whole, there were howls of protest at my remarks. This was, of course, only to be expected—who likes to be unmasked?

About eighty Members of Parliament rushed to sign a motion calling on me to resign from the Chairmanship of the Board. These were the super-democrats, the lefties, the softies and the constituency band-waggoners—a poor lot. Some of them were my former Parliamentary colleagues and I therefore knew with reasonable precision what value to put upon their expression of opinion. I certainly didn't lose any sleep on their account.

I was accused of witch-hunting, and there were frequent demands for me to produce my evidence. In fact, the evidence is there for everybody to see, and always has been. In the last unofficial strikes there was a clear relationship between the amount of Communist influence in the Area Council of the Union and the number of pits on strike in each coalfield.

In Scotland, for example, there has always been very strong Communist control of the Area Council. During the unofficial strike, many of the Scottish pits were stopped. A little further south however in Northumberland and Durham, where the influence of the late Sam Watson, continued by Charlie Pick and Alf Hesler, is still a factor, and there is hardly any Communist influence, very few pits came out. In Yorkshire we find a number of publicly acknowledged Communists, but also a great many elements from several other extreme left factions. In this coalfield the strike was widespread.

Communist influence is weak in the Midlands and very few men stopped work, but the Communists are, and have always been, powerfully represented on the South Wales Area Executive, and again the stoppage was widespread. In Lancashire, North Wales and Cumberland, there is responsible leadership and again the pits have usually gone on working.

I believe that this clear and consistent relationship between the number of pits on strike and the number of Communists in prominent positions in the local Union is much more than a coincidence. Year after year militant Areas like South Wales, Scotland and Yorkshire have caused by far the greatest loss of coal through disputes and go-slows. In every ballot on a wages offer

33

since nationalisation Scotland and South Wales have unfailingly voted against acceptance.

Can it be claimed that the Coal Board are reactionary managers in these coalfields, but reasonable everywhere else? It is the Midlands Areas of the Coal Board which have made the profits that have carried the other coalfields in almost every one of the twenty-four years since nationalisation. I think the men in those Areas and their leaders have been very tolerant indeed. In Nottinghamshire, where most of the pits are very profitable, the faceworkers have loyally accepted tiny wage increases (in some years, indeed, they had nothing at all), while the other coalfields caught up with them. This indicates a tradition of co-operation in one place, and in another a tendency to look for opportunities to kick up trouble.

There are some Communists who are far-sighted enough to see that their members will benefit only if the industry that employs them is efficient and prosperous. In the case of mining, Arthur Horner and Will Paynter were obvious examples of this. They were both trade unionists first and Communists second. Paynter remained a nominal member of the Communist Party until he retired, though in the last few years he played very little active part.

One of the most ridiculous aspects of Communists in the trade union movement has always been their enthusiasm for strikes, whereas in the Soviet Union and the Iron Curtain countries strikes are illegal. They don't seem to accept that a nationalised industry in Britain is in many ways in the same position as it would be in the Soviet Union. There are no private investors whose pockets must be filled; the whole British people benefit from an efficient and prosperous British coal industry, and although we don't have workers' control (Arthur Horner used to say that it was management's job to manage!), we do have a very high degree of involvement in the running of the industry by everybody who works in it.

Picketing of pits where men following the official leadership wanted to work in 1969 and 1970 was both violent and intimidating. Most men arrived for work in ones and twos, whereas the pickets often numbered fifty or a hundred. It takes tremendous courage for a man to force his way through a threatening line of pickets who outnumber him so dramatically. And affluence has brought new problems. The men at Woolley Colliery, near

Barnsley, for example, were afraid to go to work on the night shift because, as they claimed, their vehicles left in the car park would very likely be damaged by the pickets. The notorious flying-squads of pickets in Yorkshire were highly organised and the men were paid for the days that they spent going round the coalfields trying to stop other pits.

I realised that unless someone drew public attention to the activities of the pickets, the situation would get worse, and the risk of physical injury and perhaps damage to property would grow. After my experience in Doncaster, therefore, I said at a press conference exactly what I thought of the pickets and demonstrators and described some of the menacing tactics they had been adopting. And again came the cry for evidence.

However the evidence was produced very quickly and in a highly spectacular way before the eyes of millions of television viewers the very next day. The National Executive of the NUM held a meeting at their Euston Road Headquarters in London, and even before the meeting started the building was surrounded by a howling mob of demonstrators, many of whom were not miners, but people who travel around from one dispute to another exploiting it for their own political ends. About eighty-five delegates paid for out of Yorkshire Union funds attended as 'official' pickets, and at least two bus loads were there in addition. Scotland, Wales and Kent added their quota, and violence broke out as members of the Executive endeavoured to enter the Union offices. Between forty and fifty police were required to prevent the hostility from breaking out into greater violence. Even to get into the building at all with heavy police protection was quite an achievement for those members of the Executive who were, rightly or wrongly, thought to have moderate views. Albert Martin, the Nottinghamshire miners' leader, was kicked in the groin and injured so badly that he had to go into hospital soon afterwards for a hernia operation. Joe Gormley, of Lancashire, and now President of the Union, was another who was physically assaulted in the crush. Sid Fox from the Midlands received a black eye, and Sid Schofield, then the Vice-President, was so badly hustled that the meeting had to be deferred for a time to enable him to recover. I received a telephone call from one of the besieged leaders begging me to ask for police protection, which I did.

But these disgraceful episodes were mild indeed by comparison with what happened when the meeting finished. Lawrence Daly, the Union's General Secretary who, in the absence through illness of President Sidney Ford, was the senior national official at this time, was completely unable to leave his own office despite, again, having heavy police protection. Daly is by no means a moderate, but the demonstrators howled him down and refused to listen to his words.

Television newsreels that night had vivid film of all this and there were some dramatic pictures in the newspapers next morning of Daly, a lonely isolated figure facing out towards the road past lines of police at a mob who were behaving like animals. The whole nation, thanks to these pictures, could see that there was every justification for the criticisms I had made. Nevertheless, the demands for my evidence were still being made by the closed-mind fraternity who have no difficulty in completely ignoring events which do not suit their argument.

Picketing is perfectly justifiable where the dispute is an official one. The law allows attendance at or near a place where a person resides or works in order to inform or persuade peacefully, which means that picketing must be peaceful in character and free from both physical violence and intimidation. Picketing today seldom is, and the law is openly violated; but no action is taken. The law provides protection, as it stands at present, for picketing in both official and unofficial strikes. One of the most important measures that could be taken in the field of industrial relations would be to make picketing illegal except of workmen at their place of employment when the strike is official. This would help the official leadership, prevent the professional trouble-makers and outsiders from joining in, take the heat out of the situation, and allow logical arguments to prevail. The Government, in my opinion, would have been far wiser to introduce legislation which would have done this and to ensure that all ballots should be secret. They should insist that the ballot takes place at the man's place of employment, with the liability upon the employer to provide proper facilities.

These two legal requirements would give the rank-and-file workmen the invaluable right to make their own decisions free from the intimidation to which they are often subjected.

The miners' unofficial strike lost the men about £5 m. in wages; it lost the industry revenue to the tune of £21 m. and gave a great impetus to conversion by industry from coal to oil. The amount of coal lost in this strike, 2·8 m. tons, added to the loss of the previous October, far exceeded the amount of imports needed in 1970/71.

I am convinced, as I write this now, that if Daly had taken the opportunity when I offered it to him to recommend his Executive to use the agreed machinery of the National Reference Tribunal, he would have done much better for his members and would not have put such a strain upon the Union.

The voting for acceptance of the Board offer was 158,239 with 82,079 against in a 77 per cent poll. So, when the rank and file were at last allowed to vote on the simple issue of 'yes' or 'no' to the Board's final offer, this was the reply they gave to the Communists and the militant Left.

Even when I was about to leave the industry, and could do them no further harm, the militants vindictively tried to sour up my last pit visit as Chairman. This was to Silverdale Colliery in the Potteries, where not only had there been a long run of record-breaking outputs, but also a tremendous safety record. The colliery had won the national safety competition for its class, and also the prize for the biggest improvement in the accident rate. So keen was the will to win these awards that a surface worker who slipped and broke his leg one icy day insisted on coming to work with his limb in plaster. He didn't miss a shift, so the pit suffered no penalty, the competition being based on working days lost through accidents.

The pit had been making hefty annual profits for several years, and some of this money was ploughed back in the form of improvements to the surface buildings and amenities. A new conference room had been completed just before my visit, although work on it and other improvements had been started long before the arrangements were made for me to go there.

The pit's success and the degree of co-operation between the men and management obviously irked a few of the local agitators, only one or two of whom worked in the industry. About the same time, incidentally, Kellingley Colliery near Pontefract was heading for new European output records and the malcontents there, also hating the sight of success, tried—

but failed—to persuade the rest of their mates not to join in the bid.

A day or so before I went to Silverdale, the local evening paper carried a story that some of the men intended to boycott my visit because they resented what they described as 'bull and red-tape work done especially for the occasion'. The management and the NUM Branch President both issued statements refuting the anonymous allegations and the argument was well publicised in the newspaper.

I travelled the pit with the able young Agent Manager, John Belcher, who is the kind of man to give you confidence about the industry's future. As we went round we chatted to every man we met. I went on to Four's face in the Winghay Seam and watched the coal pouring off like a black river. Everywhere we went I got a strong handclasp and warm wishes for the future. Some of the men wanted to know if there was any chance of my changing my mind and staying on as Chairman.

Back on the surface I was presented, from 'the team at Silverdale' as the inscription said, with a beautiful figure of an old-time collier. It was the work of an eminent local free-lance artist, Peggy Davies. Done in red clay, it was exquisite; you could see the old chap's knuckles, the veins on the back of his hands and even his hair. It had been paid for by the men, as well as the management.

The controversial conference room was packed out. It was as if the men had been made all the more determined by the talk of a boycott to give me a good send-off. In fact, the Agent Manager found out afterwards that twenty-five more men had turned up for work than on any previous day. And the pit already had an excellent attendance record; absenteeism from all causes was running at about 11 per cent, compared with a national average of about 18 per cent.

So once more the agitators had completely underestimated the generosity of the miners. The publicity they had stirred up had only resulted in a warmer demonstration of affection towards me than might otherwise have been expressed.

I could not have had a more memorable last pit visit. The people at Silverdale had worked hard for their results and were enjoying their success. When I came to make my speech of thanks, I was deeply moved. In fact, I was more upset than I'd been at any time in my ten years, except for the night the Board discussed the report of the Aberfan Tribunal and I told my colleagues I had offered my resignation to the Minister.

It was my visit to Silverdale that made me realise more fully than before what I should be giving up when I came to leave the industry a few weeks later.

3

Maintaining the Dialogue

Most people thought that, as I had been for many years a professional trade union official (and before that an activist as a rank and filer, and subsequently part-time Branch official of a trade union), coupled with the fact that I had spent a lifetime in the Labour Party and ended up in the Attlee administration as Minister of Labour, my relations with the trade unions in the mining industry, and particularly the National Union of Mineworkers, would be a pushover. But this was far from the truth. I was not only right of centre, but I still retain the old-fashioned notions of men like Ernest Bevin and others who were my mentors. These were that a bargain was a bargain and must be honourably kept; that as a responsible trade union official, I accepted the complete obligation to ensure that my members carried out agreements that I had entered into on their behalf and with their complete support; that if the bargain or agreement proved to be onerous to the union members, it was necessary to negotiate changes with the employers. I also believed, as indeed I do now, that the trade unions could not operate in complete isolation from one another, but that they existed as a brotherhood that ensured that no individual group, by reason of either their strength or influence, would take action that made life more difficult for others. I believed, too, that it was not possible for a workman to behave just as an organised worker, concerned only with his own selfish wants as a worker, but that he must recognise that he was also a consumer and above all a member of the whole community.

During my Coal Board time the NUM had men like Sam Watson of Durham, Bobby Main of Northumberland, Joe Gormley from Lancashire, Will Paynter, the General Secretary, and Sidney Ford, the President. These, with others, formed just a small band of real trade unionists against the rather large group of ambitious self-seekers (whose eyes were on power, on the fees, fares and expenses book), plus the dedicated Communists, who put their Marxism before the interests of their members.

Indeed it was with those rather high motives, now of course completely out of fashion, that I first spoke at a meeting of employees of the Coal Board, including both management and workmen. This was in 1948, when I was Parliamentary Secretary to the Ministry of Fuel and Power, and I addressed the National Coal Board's summer school at Oxford. About five hundred people attended these week-long schools, which were an attempt, in the early years, to make everybody feel part of a single organisation with a single purpose. So there were miners as well as accountants, scientists and industrial relations experts, mining engineers and administrators. I pointed out that in the first year of public ownership, more than £60 million had been added to the wage bill in terms of improvements in conditions. I welcomed this and said that there were claims for still better conditions that would have some day to be met. I went on: 'It should be clearly understood, however, that the industry cannot look on the consumer as a milch cow and merely go on increasing the price of coal to cover the extra costs. That policy is quite indefensible. To win cheaper coal, whilst maintaining and even improving the standards of the workers in the industry, can come only from greater efficiency. Technical efficiency is only a matter of time.

'Greater production gives lower overhead charges and provides a surplus leading to higher standards of life of those working in the industry and lower charges to the consumer. The prosperity of the miner and all who work in the industry is dependent on the prosperity of the industry itself. And as for the nation, an abundance of cheaper coal in these immediate months and years will help to end our shortages and help to secure a fuller and better life for the whole community.' This was a philosophy, first expressed one year after nationalisation, from which I never departed.

Of course, coal had been scarce throughout the war and continued to be so for many years after I made that speech. In fact, the industry really didn't have to start selling its products hard until about ten years later.

And yet, when I came to be Chairman of the Board thirteen years after I made that speech, there was still a feeling that the world owed the industry a living. Marketing was an art which had long been forgotten, or at least had been dormant and unused for about twenty years. The theme of sharing the improvements and technical progress between the miners and customers was one which I had to go on arguing in many of my annual speeches to the National Union of Mineworkers. Someone said recently, and with great wisdom, that nothing can withstand an idea whose time has come. It is also true that human beings find it very, very difficult to abandon ideas that are no longer valid. Human nature still insists on learning most of its lessons the hard way.

There are a lot of people with stars in their eyes who argue that what we must do is see that every worker gets job satisfaction. This is just not possible in the case of many thousands of jobs in industry. Where the types of work do not permit it, men look elsewhere for a sense of fulfilment and offset the tedium of their day-to-day jobs by throwing themselves into all sorts of activities, such as local government, religious organisations and trade unions.

But the really important thing is to get the rewards right. I have long believed, and a few months after my arrival at the Coal Board made a public suggestion, that the sensible thing was not to wait for wage applications at all. Management and the unions should agree in a particular industry that every year they would sit down, look at what the industry had paid in the past year, say roughly what might be done in the following year, define what level of production might be achieved (taking into account the new machines that might be available), apply work study methods and then agree that next year's wages could be raised by X per cent. This would provide a basis for better earnings, increased productivity, better standards of living, and be a sensible and civilised method, provided that both sides entered into it with complete honesty and fairness.

Though we achieved many of my other ambitions in industrial relations, we didn't get as far as this. I was too much of a revolutionary in these matters for many of the militants who took pride in their radical approach, but in fact were incapable of even beginning to think of changing the way in which wages were negotiated. I knew from the beginning that mechanisation was going to affect relations with the men, but it was essential to the industry's whole future prosperity. Nevertheless, all the benefits could not go to the worker—that was syndicalism and I was far from being a syndicalist. Quite a lot of the miners' leaders, particularly at Area and pit level, undoubtedly were.

There was another basic factor in my approach to industrial relations: the need to achieve a balance between my responsibilities to the people of the industry on the one hand, and to the nation as a whole on the other. As a member of the Labour Government that had nationalised the fuel and transport industries, I was always conscious of the possible danger that the workers in those industries—and especially in coal—having had a raw deal in the past, would demand that all the benefits from greater efficiency should go to improving their conditions. There was an obvious danger that the consumer might be forgotten. In fact, of course, the industries were nationalised for the benefit of the community, not just for the people who worked in them, urgently though their conditions needed improving. This is one of the extra complications for management in the nationalised industries.

I had been for four years the junior Minister at the Ministry of Fuel and Power under Hugh Gaitskell, when I visited Washington. Of course, as I represented in Parliament a mining constituency, I was interested in American mining methods. I therefore made arrangements to see John L. Lewis, the President of the United Mineworkers of America. I knew that he was vigorous in his enthusiasm for mechanisation and I was anxious to discover his views upon the social consequences. During our conversation I said that surely the great problem about mechanisation was the impact it made on the lives of people. All sensible people welcomed mechanisation, for the machines could do the hard labour that man had been forced to accept for centuries. Nevertheless, these technical changes could obviously cause misery. I argued with Lewis that if they were not planned in association with other outlets for the labour no longer required, there would be great domestic and social problems. To clinch this point I said: 'Let me put a hypothetical case to you. Suppose a mine operator said to you: "Mr Lewis, I have some new machines that will be put into the mine next month and instead of employing 2,000 men I shall want only 500 to turn out the same amount of coal." What would your attitude be?'

Lewis answered: 'I would say to the mine operator, O.K., get on with it.' When I asked him about the 1,500 men who would have to go on the road he just shook that great shaggy head of his and laughed into my incredulous face, saying with a roar: 'The great expanding American economy will take care of all those boys, and in the meantime I shall want another dollar or so a shift for those that remain.'

I was not wholly convinced and since then I have seen the American ghost towns and the flotsam and jetsam of humanity that has clung on, existing from the soup kitchen. So did John F. Kennedy when in 1960, during the Presidential Election, he toured the Appalachians. Kennedy could scarcely bring himself to believe that human beings were forced to eat and live off cans of dry relief rations. I recounted my conversation with John L. Lewis in my last address to the NUM annual conference held in Douglas in July 1970. Naturally, I made it clear that I was on Kennedy's side and not John L. Lewis's.

Even before my predecessor, Jim Bowman, gave up the Chairmanship, I was involved in my first round of wage negotiations with the National Union of Mineworkers. These started in the middle of January 1961, and there was obviously a risk that they would not be completed before Bowman left. This was the first time I had faced the NUM's negotiators across the table. They were led by Sid Ford, who had recently been elected National President, and Will Paynter, the General Secretary, who was then a member of the Communist Party. I grew to admire and respect both these men, though there was a gulf between them politically and theirs was an uneasy partnership. But Ford and Paynter were united in one thing: a determination to secure the commercial future of the mining industry and thus of the men employed in it. Both were realistic, both were honest.

Sid Ford had never worked in a pit and had gained his union experience in the London office of the NUM. This must have been a handicap to him, but he made up for any lack of technical knowledge about the industry by his courage and determination. In his early years as President he had the tremendous help of Sam Watson, the Durham miners' leader, and later on, even without this great little man's help, Ford resisted successfully attempts by the extremists to take over the Union. It was a great tragedy and a great loss to the trade union movement when he was forced to retire early for reasons of health.

Will Paynter, like Arthur Horner his predecessor as General Secretary, was a product of the Rhondda Valley. A fiery and effective orator, he had also been involved on the Republican side in the Spanish Civil War. As a negotiator, he was without equal. His mind was logical and his presentation of a case always cogent and powerful. Like Ford, he had tremendous moral courage and was quite prepared to face unpopularity among his members for what he thought was right. The progress that was made towards getting rid of piece-work in the mining industry will, I am sure, be regarded as historically important; and a great share of the credit is due to Will Paynter.

My political associations in Northumberland had also inevitably brought me into close personal contact with NUM leaders on both sides of the Tyne, and nobody could have lived and worked in the House of Commons as long as I had without having a profound consciousness of the political strength and weaknesses of the National Union of Mineworkers' Parliamentary group. To that extent, I started off with something of an advantage.

Every year since nationalisation, the National Union of Mineworkers have invited the Chairman of the Coal Board to address their annual conference. It is only fair to say that not all of the Union leaders approve of this practice, and they made this abundantly clear from time to time. My first experience of this occasion, which is probably unparalleled in any other industry, was in July 1961, when the conference was held at Rothesay. I was quite surprised, when I got to my feet, to see some of the Scottish and South Wales delegates ostentatiously reading newspapers throughout the whole of my speech. They turned the pages over with as much paper crackle as possible. What surprised me was that they should think that their rudeness should have any impact upon me at all. I regarded this sort of action, as I regarded aimless demonstrations subsequently, as churlish childishness. Nevertheless, these opportunities for a direct expression of view to the entire leadership of the Union were extremely valuable and, even on my debut, I didn't hesitate to speak bluntly about the need to maintain price stability against the competition then mainly coming from oil. I also spoke very frankly about the £40 million which had been lost because of absenteeism and unofficial stoppages during the previous year. I pointed out that this would have been enough to pay for an increase of 10s. 0d. a week to all the workers in the industry.

But I always took the chance to express positive ideas, too. For example, as early as the 1961 conference, I was already talking about using the new machines to earn enough profit for miners to work a thirty-two-hour week and to retire at the age of sixty. Remembering that the 'secret decentralisation plan' had so monopolised the interest of everybody at the previous year's conference, I set out to kill the suggestion that this would involve any return to pre-war district wage agreements. However I told my audience frankly that decentralisation of administration was something that I fully believed in. I explained that I was talking about normal day-to-day administrative decisions; planning of investment, planning of production, marketing policy, and national negotiations on wages and conditions all being left at the centre. But for the rest, we would delegate to the manager on the spot as much authority as possible. I reckon my forthrightness was appreciated because I got a warm and prolonged burst of applause when I sat down. But not from the closed minds. They put down their newspapers and sat on their hands.

During that summer of 1961 Selwyn Lloyd, the Chancellor of the Exchequer, imposed his historic wages freeze. The NUM had submitted a claim and in September they agreed to defer it. They said they were doing this, not because of the pay pause, but because they had been impressed by the better results in the industry. They thought that they would have a better hope of success if they allowed the current rising trends to go on a bit longer. A couple of months later there was a great row in the House of Commons because the Minister of Power, Richard Wood, gave the impression that the Government had warned me not to defy the pay pause in the resumed wage negotiations. Obviously, if the idea that I was carrying out Government policy gained credence, my authority in the wages talks would disappear. And I had certainly had no instructions from the Government. So I quickly issued a statement saying that of course I knew of, and thoroughly understood, the Government's views on the pay pause, as did every other employer and trade union leader in the country. But I had received no instructions from the Minister or any other member of the Government. I regarded myself as being perfectly free and uncommitted in my approach to the current negotiations. I further declared that if I had been instructed as to my course of action, then I would have had to inform the Union to transfer their negotiations to the Government. And that remained my attitude throughout my Coal Board time.

The round of wage negotiations in the early months of 1962 ended with the Coal Board offering 7s. 6d. a week increase for day-wage men. The final settlement I negotiated gave increases, in some cases, of £3. The two figures show vividly the effect of wage inflation in that period of eight and a half years.

When the ballot was finally held, the offer was accepted, although Scotland and South Wales voted for outright rejection. In fact, these Areas consistently voted against every settlement that was negotiated during my ten years. And yet these are the coalfields with the lowest productivity and, generally speaking, the worst strike records. The fact that they were able to enjoy wage increases at all was due to profitable pits in other coalfields. Since nationalisation, the Scottish mining industry has lost about £190 m. after meeting interest charges, despite the benefits of a differential in price over most other coalfields. South Wales has in the same period lost about £145 m. On the other hand, the East Midlands pits have to their credit an accumulated surplus after interest of about £265 m. Without these excellent results, the Scottish and South Wales coalfields could simply not have survived.

The resentment within the Union ranks on the part of the Nottinghamshire miners is a considerable force even today. The East Midlands miners, though earning such wonderful profits for the industry, have rarely felt able to indulge themselves in the luxury of a strike. In South Wales and Scotland, on the other hand, the losses have rarely inhibited industrial action over a great many issues that would have been settled sensibly in the Midlands. Yet the coal in South Wales is of superb quality: the anthracite is some of the finest in the world, there is good coking coal which is in extremely strong demand, and the dry steam coals are an essential part in the movement towards clean air. The Union leadership in these two coalfields has a lot to answer for.

East Midlands coal, a much more difficult marketing proposition, was only saleable because of the productivity that came from the excellent relations between the men and management and kept down its cost. If the industrial behaviour of the East Midlands miners had been as truculent as that of South Wales or Scotland, the coalfield could not have prospered.

I very much valued the opportunity of speaking at those annual conferences of the Mineworkers, and I knew that despite the hostile attitude of the militants the bulk of the delegates were pleased to listen. Equally I knew that they did not want to be given a lot of 'soft soap' but expected to hear a frank exposition of the current mining situation, even if many of the things that I said were not wholly palatable.

It was a splendid chance to speak freely which I always took; I saw no point in travelling each year to the conference simply to exchange a few polite platitudes, and the delegates would have been disappointed if I had done so. I always tried to take a pragmatic approach and one which represented my own personal considered views. On one occasion I spent some time attacking a pamphlet *A Plan for Miners*, issued by some groups within the Union. This publication had criticised the Coal Board's accounts and the amount of interest being paid. Its purpose, of course, was to show that there was, or could be, enough money in the kitty to meet in full the miners' wage demands. This was untrue and I was determined to resist this and other attempts to mislead the miners—as indeed I did all the time I was with the Board. The truth was always twisted by the knavery of the militants to make a trap for the thousands of good, hard-working men in the industry. They did great damage with this sort of thing, not only to the men they claimed to lead to a better life, but to the industry which they served, and as a result constantly reduced the ability of the Board to provide the improvements they never ceased to demand.

In my speech, which was punctuated by the customary boos from the usual delegates representing the usual coalfields, I showed what a small proportion of our interest payments went to the former owners. This was a contradiction of the impression that the pamphleteers were trying to create—that the £41½ million interest payments during the 1963/64 fiscal year were going to the old coal owners. Any adverse reference to the old coal owners was usually good fodder for the troops. But this impression was soon negatived as I exposed the facts. The militants always grossly underestimated the intelligence of their own colleagues in the pits, or if they themselves did not underestimate the mineworkers' intelligence, then the academics from the universities who mainly wrote this brand of pamphlet for the Left undoubtedly did. I never made this mistake. Frankness and honesty always pay, not only with miners, but with workpeople generally, so I broke down the figures, leaving the lads to add it up.

The stocks of coal had to be financed—that was £4 m.; the Board deficit had to be financed and that was another £1 m.; since nationalisation the Board had borrowed millions of pounds for new machinery, new pits, better safety and health measures, new houses for miners and so on. The interest on this was £32½ m. Only £4 m. in fact represented interest paid to the old owners. I reminded them that no one lent money without interest; they were nearly all members of co-operative societies and received interest on their modest share capital. Then I hammered home the final, and perhaps most telling fact, which was that the Miners' Pension Fund, which collected interest on its investments, had a greater income from that source than from the miners' own weekly contributions. The canard of interest was finally exploded and we never heard of it again.

One of the sponsors of *A Plan for Miners* was Will Whitehead, then a Communist and President of the South Wales Area. Whitehead went to the rostrum soon after I had finished speaking and said he had not joined in the hand-clapping at the end of my speech. He resented the linking of the pamphlet with the Communist Party, which he said was as untrue as to suggest that the speech of the President, Sid Ford, had been written in Hobart House. He said he still did not accept my figures. However there was no lasting ill-will between Whitehead and me, and when he gave up his Union job some time later, and many of his former friends turned against him, I helped to find him work elsewhere.

There was a continuing tendency for some of the miners' leaders constantly to look to the Government to save the industry from contraction, and at the same time to ensure substantial improvements in wages and conditions. This attitude did not further the interests of the miners, because it took their eyes off the economic ball. Efficiency alone determined the size of the industry and consequently the number of people it could employ. They really believed that under a Labour Government life would be different, and it was some time before the disillusionment came. I took the opportunity of the annual conference speech time and time again to try to show the realities of life under any Government. The message was a simple one. If the mining industry was to be successful and prosperous, then it was necessary to utilise to the limit the resources that we had at our disposal and over which we had complete control. Only when that had been done could we expect the Government or anyone else to lend a hand during the difficult transitional period through which we were passing. Reading through these speeches again, there is nothing that I uttered then that I would want to alter in the light of subsequent events.

In 1968, the year after the White Paper on Fuel Policy (about which I shall write later), I tried to bring the Union face to face with reality. I told them that they had all thought that with a Labour Government life would be easy for the mining industry. Now they knew differently. 'You will live by pulling yourselves up by your own boot straps. You cannot look to Governments to save you any more,' I added.

By then the bitter lesson had been learned, but it had taken a long time. After about twenty years, during which the policy had been coal at any price, the feeling had existed that the country owed the mining industry a living because of its past neglect. For my part, I repeatedly tried to show that in the end it was the customer who decided the size of the industry. Once it was cheaper or more convenient to use another fuel, sentiment would count for nothing.

To be fair to the NUM, they had held back from ruthlessly exploiting their position when coal had been desperately scarce, especially during the time of the post-war Labour Government. Quite rightly Arthur Horner, Paynter's predecessor, used to say in those days that if the Union had asked for the moon, the Government would have had to get it for them. During my ten years the position was very different. We had to fight for every ton of business and it was this that I wanted the Union to understand. The militant section of the NUM did irreparable harm to the industry and the men they were supposed to be fighting for. What they always forgot to take into account was the effect of their disruptive tactics upon the customer. Strikes of any duration always did two things: first, we lost good men who left the pits rather than tolerate the strikes; and second, we lost customers. We had a difficult enough task to sell coal at that time in competition with cheap oil; if we then added a risk ingredient, the industrialist tended to move away to the more reliable oil supply. I am quite certain that a good deal of pressure to move to nuclear power was as much to do with escaping from the dependence upon the coal mining industry and the railways, which were needed to move vast quantities of coal to the power stations, as the desire to move into a new technology.

In the spring of 1968 industrial relations in the industry were in a sensitive state. Not surprisingly the miners' leaders, inspired by the bleak picture of the industry's prospects given in the Government's Fuel Policy statement, were inclined to adopt tactics of despair. A call for industrial action as a protest against the Government's policy was defeated by only 14 votes to 10 at a meeting of the Union Executive. Sir Sidney Ford, President of the Union, reporting the Executive's decision afterwards, said that the majority had considered industrial action would create untold hardship without any hope of achieving any improvement for the great mass of workers in the industry. Indeed the likelihood would be that they would finish up after any stoppage with an even smaller industry. It was certainly an unhappy time for people like him and dozens of other loyal Labour Party members in the NUM. Only this loyalty and the fact that the Coal Board were publicly and vigorously fighting the battle for miners' jobs prevented severe industrial action at this time. Nevertheless the battle within the Union went on, and this internecine warfare made great difficulties for management as well as creating bitterness and confusion amongst the men themselves. In 1969 war broke out, as I have described in the previous chapter.

When I gave my 1970 address to the Union's annual conference in Douglas, the delegates had already adopted a resolution calling for an official strike unless the Board conceded in full the claim for a £5-a-week increase which they were putting forward. I knew that there was real danger of another unofficial stoppage if we failed to get a settlement on this huge claim which, I told them, would cost the industry £75 m. in a full year. So I set out bluntly to make the Union see the damage that had been done to the industry and to their members by the last stoppage. This part of my speech provoked more fury from the militants than from any of the other ten addresses I'd given at the annual conferences—and some of those had been pretty controversial too.

I said first that the activities of the militants and the unofficial strikers did not contribute to the solution of the task which faced us. 'Far from them securing great wage and social advances, they are like Samson who used his great strength without much intelligence, and pulled down the pillars of the temple, destroying the temple and himself in the process.' The men who took part in the strike had lost £4 m. in wages. The Coal Board had suffered a loss of income of £13 m. Customers decided to turn to other fuels upon which they could rely, and the Board had been compelled to go for increased prices. After the uproar had subsided, I went on relentlessly: 'This unofficial strike bled the industry by self-inflicted wounds. It broke solemn agreements. It reduced the power and authority of the Union. It was an act of anarchy.' I told them that the leaders of the unofficial strike were quite wrong in their claim that militant action paid. When the pandemonium died down for the second time, I pressed on, ruthlessly rubbing in the facts. The crucial point was that we had already decided to provide a minimum wage of £15 a week and to reduce the hours of the surface worker before the strike had begun. 'The strike achieved nothing,' I said, 'except damage to the men and damage to the industry. I therefore tell you this with great sincerity. Action on these lines is what brings this industry to its knees: and the worst sufferers will be the men, their wives and children. This is the negative approach to high earnings. The positive approach is to make the machines do the work that they have been designed for. This is the only way to ensure long-term benefits for the men who work in mining. There is in fact no other way. Only coal coming off the face rings the bell in the Board's cash register.'

The majority of the people in the conference hall had been opposed to the strike and they realised that I was helping them to impose their authority on the turbulent minority. When I sat down at the end of my speech there was a great burst of applause which drowned the howling of the one or two militants who had not given up their attempts to shout me down. So that the contents of my speech, and particularly this passage, should be widely known throughout the industry, we arranged to post a copy to the home of every man in the industry. Naturally, the militants were not too pleased about this either.

The tactics of the militants are usually the tactics of the bully—loud-mouthed, heckling demonstrations designed to intimidate. Indeed, the tactics of dictators throughout the centuries. Some of these people incidentally are to be found on platforms for peace in Viet-Nam, peace in the Middle East, better understandings between East and West, peaceful negotiations rather than war—wonderful sentiments for international behaviour. But their hypocrisy was exposed by their constant desire and urging for war in industrial relations at home.

I had many, many experiences of this kind when attempting to put across the simple economic facts, and I always expected and got a stormy reception. I remember addressing a meeting of the Scottish TUC at a time when a negotiation on a wage claim was proceeding. I knew that the delegates from the NUM in my audience would resent the invitation to me to address the conference. So I took the initiative by referring to this early in my speech and, sure enough, they fell for the ploy. Members of the miners' delegation stamped their feet and made it quite clear that they did object to my being called a fraternal delegate. This of course immediately put them 'in wrong' with the rest of the massive audience, who wanted to hear what I had to say before condemning or praising me. Instantly, I turned on the hecklers and said: 'Despite the bends and the twists and the distortions that naturally come about in arguments and discussions, the miners never had a better friend than Alf Robens.' I went on to say that I was running the industry according to the Act of Parliament passed by a Labour Government of which I had been a member. The Act would have to be changed if that was wrong. I hammered home my unchanging argument that the benefits from greater efficiency and productivity must be split three ways —more money and better conditions for the workers; more money for more machines to get still higher productivity; and price stability, and even price reductions, for the consumers. I reminded the miners that, because of their concessionary coal, they were insulated from the effect of increased prices in the domestic market.

I received a loud ovation when I sat down, though not from the Scottish miners, who kept their hands in their pockets as usual.

After the session as I walked out a number of the Scottish miners' delegation tried to hustle me. I stood my ground and made it clear that if they wanted trouble I was ready for it. By that time TUC officials, seeing what was happening, came to my rescue and as the 'democratic militants' departed they threatened: 'You'll get your answer in the coalfields.' What a laugh! Hitler had nothing on these boys in the use of threats. The Scottish NUM leaders still haven't forgiven me for my remarks more than seven years ago. But that hasn't stopped their members giving me the usual friendly welcome when I visit their pits.

Demonstrations were the order of the day about almost every grievance. They took the place of rational discussion and were another vehicle for the militants, who took full advantage of it. I never minded the demonstrations because one undoubted advantage came from them. They let off steam. They never prevented my visits to the pits and the coalfields, and so I was on the receiving end of demonstrations about wages, pit closures, rents of Coal Board houses, temporary loss of amenities through opencast coal operation, and anything anyone could think up. Some were angry, some good-humoured, some well-organised, some pathetic. I never tried to avoid them, although many times it was suggested to me, for example, that if I entered a hall where I was about to give a speech by the back entrance, I need not face the people with placards and loud voices.

However I made a point of taking the initiative and always going over to ask what the demonstrators wanted to say to me. This meant that they had to listen to me explaining why the Board had taken a certain decision and not another. They really provided me with a good open-air meeting, to which I was not unaccustomed. I could not claim, though, that I made a great many converts to my point of view. After all, if a man is waving a banner and publicly committing himself to certain views he is not likely quietly to fold up the banner and go away because the Chairman of the Board has persuaded him to a different point of view. But at least no one can ever claim that I dodged talking to them. A kind of dialogue was maintained, even at the height of the most bitter dispute.

One of these confrontations was quite spectacular. It took place in January 1965 at Seaham, County Durham. I was driving up to the colliery with Bill Reid, the Divisional Chairman, when we saw a crowd of demonstrators, one of whom was shaking his fist at me. Reid urged the driver on into the pit yard, towards the training centre which I was going to visit. However I ordered the car to stop, got out, and walked straight into the crowd of about eighty men whom I found were demonstrating about the proposed closure of the nearby Lambton D pit. I shoved my face straight into the face of the fist brandisher and demanded to know why he was menacing me. Not surprisingly, this took the embarrassed lad aback, and he slipped away into the crowd, saying he did not wish to be offensive.

I had acted on an impulse in jumping out of the car and didn't exactly know what would happen. But I was damned if I was going to take fist brandishing lightly at a time when I was fighting the Labour Government tooth and nail in defence of the miners and their jobs. The unfairness of the situation was too much for me and I couldn't help reacting. Incidentally, a press cameraman who was on the spot later won an award for his picture of this tense moment. It could have been ugly, but it wasn't. The colliers and I discussed their problem for about half an hour, with the crowd hanging on every word. If anyone tried to interrupt he was hushed by his neighbours. At the end of it all, the leaders expressed their thanks for the opportunity of discussing the situation, formed up their contingent in well-disciplined order and quietly marched away. What could have been a very nasty incident turned out to be a useful confrontation.

Another example of demonstrations was during the campaign over the Union's 1970 wage claim. During the argument over this claim, I intensified my pit visits to go to as many as possible. Quite often, I did an underground visit in the morning with a trip round the surface of a nearby colliery the same afternoon. One day I was underground at Hucknall Colliery in Nottinghamshire and toured the Newstead surface. The Union in Nottinghamshire had banned overtime as a protest, but this had little effect. When we got to the coalface at Hucknall it was the men's 'snap' time (the twenty minutes they are allowed to eat their food). Usually they would work straight through but, because of the overtime ban, they apparently felt obliged to stop work. However after five minutes or so, they got bored with this and started to produce coal at a tremendous rate.

I knew there was going to be trouble at Newstead: the pit had a brilliant record. Wilfrid Miron, the Regional Chairman, used to describe this and other Nottinghamshire pits as 'the Grenadier Guards of the industry'. The problem was that Newstead was running out of coal and, unless a lot of money was spent on developing into new reserves, would have to close in five years or so. When we got to a point a couple of hundred yards from the pit yard, we met two separate demonstrations—one of women. The men's affair was exceptionally well organised. There must have been getting on for two dozen banners, most of them recounting the pit's achievements over the previous fifteen years. But one was very different. It said:

> *Ode to Lord Robens*
> Hills of Annesley, bleak and barren,
> Where my thoughtless childhood stayed—
> And proud Newstead, record breaking,
> Will you once more be betrayed?
>
> (Apologies to Lord Myron)

This was an adaptation of a fragment of verse by Byron, who lived at nearby Newstead Abbey. The allusion to Wilfrid Miron was neat, too. The demonstration went very much according to form. I got out of the car and walked through the demonstrators, looking at the banners; the leaders of the demonstration gathered round to talk to me and the Press and TV men followed and scrummaged round. I had a strenuous meeting with the leaders of the unions at the pit, at which they said they had been led up the garden path in the past, that they had earned some capital investment in the colliery, and that if new reserves were not allocated to Newstead, they were perfectly capable of working so that the existing reserves would last ten years at break-even point instead of five years at a profit of £1 a ton. I promised them a final decision before Christmas: in fact, the Newstead men got their Christmas present of another half-dozen or so years of life for the pit.

And this is where the demonstration by the women links with the men's. The wives came to protest about the condition of Newstead Old Village, a group of back-to-back terrace houses, still with outside lavatories and all the usual characteristics of an old-fashioned slum. I talked to the women as I went into the pit yard, and also visited the houses afterwards. I quickly got right down to the details, saying that the outside lavatories should not be demolished when the houses themselves were renovated, but should rather be turned into lock-up sheds for bicycles and so on. Because this was part of a package: if it was worth spending money on Newstead Pit because of the high standards and high performance of the men, it was worth also spending some money on their houses. It is possible to find enthusiasm and co-operation among slum-dwellers, but it is a mean and short-sighted employer who thinks he can rely on these virtues if he continues to do nothing about the living conditions.

It was not often that the NUM organised a national demonstration, other than a conference, but their final protest about the Labour Government's refusal to implement their pre-election promises took the form of a rather sad march through London in February 1966. Carrying their banners and led by a pipe band, about one thousand men walked from Hyde Park to the House of Commons, where their leaders met Fred Lee, who was then the Minister of Power. At the end of the day many of those present must finally have realised that their arguments were falling on deaf ears. Neither with the Labour Government, the Labour Party, or the TUC were the miners the power and the influence that they once had been. Certainly, Harold Wilson's appointment of Fred Lee was not one to reassure the miners about their importance to the Labour Party.

In the first two years of his administration Wilson would have been much better advised to have given the job of Minister of Power to a miner. But not the miner he eventually appointed— Roy Mason. As I shall show later, Mason was responsible for the biggest single blow ever inflicted on the miners' morale.

My attempts to sustain an effective dialogue with the miners were based on candour and a genuine concern for them and their families. They were all the more necessary because of the treatment the industry received at the hands of the Labour Government.

4

Battling for Business

All through the war, and for twelve years afterwards, coal was 'allocated', and selling in the true sense of the word did not arise. The immediate post-war years were the most difficult; the Continental countries were desperately short of fuel and their own coal industries were in no position to supply all their needs. I remember Ernest Bevin, when he was Foreign Secretary, sending for me on his return from a high-level conference in Europe. I was Parliamentary Secretary at the Ministry of Fuel and Power at that time. He explained the problems with which he was faced and the difficulties of the four-power administration. It was clear that he had had a rough time. 'If I had a few million tons of coal to offer Europe,' he said, 'I would be able to negotiate from strength. What can be done to provide coal to the Continent?'

'There is no difficulty about supplying one, two, three or even more million tons of exports,' I said. 'The real problem is to determine which British customers, who also want coal, must be starved in order to supply the Continent. There is not sufficient coal to meet all the demands and whilst every effort, including Saturday morning working, is being made to increase output, there is no possibility of stepping up exports without closing some of British industry down.'

In the event, some exports to the Continent were shipped, but only at the expense of buying American and Polish coal soon afterwards at high prices, the loss falling upon the Coal Board to the tune of £70 m. over the years. When the financial results of the Coal Board were being discussed later, this fact was conveniently forgotten.

Plans for the construction of oil refineries in the United Kingdom were got under way and in the middle 1950s, as the refineries came on stream, the fuel shortage came to an abrupt end; within two years vast undistributed stocks of coal (about 37½ m. tons) were lying on the ground.

The marketeers in the Coal Board had to start all over again to go out and sell coal, and it certainly took some doing. Oil was now unbelievably cheap, so that the market for coal was slipping away more rapidly than output could be brought down. Our examination of the market clearly revealed our problem. We had six areas of sales: electricity, steel, railways (which were going over to diesels), private industry, gas (which was increasingly being made from oil products) and the domestic market.

Only one of these—electricity—was a growth market. All the rest were certain to decline.

Our task was perfectly clear; we had to ensure, as far as we were able, that we maximised the use of coal in power stations and at the same time put up the most massive resistance to the speed of run-down in the other markets. And this is what we did. The nuclear power programme for electricity was a serious blow and, as I shall show, a financially disastrous policy for the nation. Gas from coal was more expensive than liquified methane or gas from oil, using naphtha as a feed stock. The domestic market was hit by the slum-clearance programme, which meant that high-rise blocks of flats were built using oil-fired boilers and gas for heating and hot water, in place of thousands of houses burning coal. Our market was naturally diminished. Then the Clean Air Act added its stimulus to the use of fuels other than coal and, whilst it was a very necessary piece of legislation, it obviously made our selling job infinitely more difficult.

It was abundantly clear that, unless the runaway from coal could be substantially checked, the resultant social consequences of the displacement of miners from their jobs would be catastrophic. This was a fact totally ignored by the leaders of the other coal-using nationalised industries and only lightly considered by Governments. Labour leaders subsequently paid much more attention to the problems of redundancy in connection with the Upper Clyde Shipbuilders than they did as Ministers when we had the problem of many thousand redundancies in mining.

The brake on pit closures and redundancies could only be applied by allowing the accumulation of large stocks, which had already been done to the maximum, and by holding the market to the best of our ability. It was a tough assignment, but we had a good marketing organisation, well led by Frank Wilkinson, who had been selling coal all his life and, after two post-war spells sorting out fuel distribution in Germany, was Marketing Member of the Board. He was ably assisted by Derek Ezra, now the

Chairman, then Director-General of Marketing. The campaign was well conceived and well organised, and we changed gear from allocation to high-powered selling.

In this we were highly successful, so much so that, although I personally did not sell a lump of coal, I was nominated joint 'Marketing Man of 1964' by the Institute of Marketing. The award was shared with Monty I. Prichard, Chairman of the Perkins Group of Diesel engine manufacturers.

I never hesitated to involve myself personally in our sales promotion activities, although I realised that some of my colleagues and some people in managerial posts outside the industry felt that this was undignified for the Chairman of the Board. But I took part in television commercials and also in direct mail activities designed to sell householders the idea of solid fuel central heating. Later on, we relied to quite a considerable extent on direct mail which enabled us to reach householders in just those social classes and income groups where we knew we had our strongest appeal. The booklets we sent out were accompanied by a letter from me with a photograph. The results were excellent and we found we got our inquiries at a cost that was very attractive by comparison with more conventional selling methods. However, I would not allow our people to adopt foot-in-the-door methods. No salesman called on a householder without an invitation. Of course, we made it easy for the householder to extend the invitation, because we put reply-paid cards in our mailing. But the point was that no one arrived unexpectedly to pester householders who were not interested. And of course I only involved myself in advertising that was skilfully and well done. I am sure that it showed the people in the industry that the gaffer was quite prepared to get stuck in to try to win markets for our products.

The domestic heating sales campaign, begun in my first year of office, gained quick initial success. The oil companies had been the first to promote central heating, but people whose interest was aroused by their advertising found, when they got the estimates for an installation, that it was going to be too expensive for them. So they looked at systems based on other fuels. We advertised on a much more modest scale, but walked off with a far bigger share of the new business because solid fuel was cheaper. In fact, in 1962, market research showed that solid fuel had captured more business than all the other fuels put together. During the year the central heating market had grown very rapidly, and 63 per cent of the new installations were ours. Gas and electricity had 11 per cent each, and oil had 9 per cent.

Over the years people became more affluent, and the greater convenience that gas and electricity were able to offer by comparison with solid fuel made them more popular; we gradually dropped to third place. Oil is more attractive in the bigger systems and has only a minority appeal. Its market share has remained at about 9 per cent.

The rapid improvement about nine years ago in the design, efficiency and appearance of solid-fuel equipment was a credit to the manufacturers. The good-looking roomheaters with back boilers for central heating proved tremendously popular. People still enjoy the sight of a 'Living Fire', as we styled it in our advertising, but our competitors were able to concentrate on greater convenience. So the battle went on.

Our domestic market was very important to us, as it represented about 30 million tons a year of profitable business. The gradual designation of smokeless zones however meant that more and more manufactured fuels would be required. It was also clear from our own experience as producers of smokeless fuels that a fantastic amount of capital would be needed to build smokeless fuel plants and that this would not be forthcoming. The Government were hardly likely to provide capital for solid smokeless fuel when they had already provided huge amounts to distribute North Sea gas. We would need to find other ways and means if the domestic market was to be held. So we turned back to ideas and proposals that had, some years before, been virtually abandoned; these were to design and produce an appliance which would burn raw coal smokelessly.

Our research scientists engaged themselves in this work and eventually made an appliance which did the trick. Our problems were not solved entirely. What the scientists had produced were 'one-offs' and these had to be turned into something suitable for mass production. We worked, then, with a manufacturer and we arranged for one thousand of these appliances to be made. Obviously these had to be given market trials. They had to be submitted to test in the home by householders whose uses would be widely different.

The decision was made to take advantage of two housing estates owned by the Coal Board, one at Brimington, near Chesterfield, and one near Nottingham, which were to come under a smoke-less-zone order. Installation went ahead and then our problems really started. With every new thing that is produced or designed it is never the principle that is found to have failed, but the bits and pieces which for a variety of reasons go awry. Some of our tenants were angry and unhelpful, some pleased and co-operative. I was eternally grateful to the women's organisation on one estate who went out of their way to make the scheme a success, as compared to the tenants' association, whose main purpose seemed to be bellyaching all the time.

Despite the problems and, I'm afraid, the inconveniences to the tenants, the trials were highly successful. We were able to identify weaknesses and difficulties as they were experienced and to put them right. Some of the problems related to the type of house. What clearly emerged from the trials was that the system that had been designed by our scientists to burn raw bituminous coal smokelessly was correct.

Having established that, we offered the research results to two of the largest appliance manufacturers in the country, who are now producing bigger versions of these appliances. The type of coal used is extremely cheap, giving a big saving in fuel costs to the housewife. In fact, what the domestic consumer will have as a result of this successful research and the patience of our tenants, is a living fire, domestic hot water, and controlled heating at a running cost well below gas, electricity or oil. This will not only slow down the loss of the domestic market, but in due course should turn the tide and, with the large housing pro-grammes over the years ahead, even increase this market for coal.

The beneficial effects of competition between the different fuel interests have certainly been very evident in domestic heating. The rise in the number of homes having central heating has been steady, although there are still many millions of houses inade-quately heated. The improvement in design and efficiency of the equipment has been spectacular. The campaigns promoted by the various industries have been an example of the beneficial effects of advertising. People's homes have been made pleasanter places to be in. And I believe they have become healthier as a result—

despite the old British feeling that there is something immoral about having a warm bedroom.

Nevertheless, there has been some unnecessary heavy expenditure on advertising by the fuel industries. I tried to persuade the other nationalised industries—gas and electricity—to join in an intensive co-operative campaign that would have promoted the basic idea of central heating. We would then have provided inquirers with information about the various kinds of heating and the different fuels they could choose from. I also advocated the setting up of a nationwide independent objective service that would have advised householders on the type of central heating most suitable for their purpose. I suppose it was because I was involved in the original legislation concerning nationalisation that I was anxious about the impact of public ownership upon the consumer. Whilst in most things the various fuel industries had to develop along their own way, yet there were many possibilities of commercial co-operation of benefit to the consumer that were ignored. It would have been comparatively simple and far less costly to have advertised central heating as such on a co-operative basis and then to have helped the consumer to select the right kind of fuel for his particular circumstances. Instead of that, millions of pounds were spent in competitive duplication.

I tried to interest the chairmen of the other nationalised fuel industries in this scheme but got nowhere at all. No doubt they thought I was trying to put one over on them. During the time that Richard Marsh was Minister of Fuel and Power, his department were interested enough in the idea to conduct some inquiries, but nothing came of it all.

But all the time we realised that the price of coal or solid smokeless fuel to the consumer was perhaps the most vital element in selling. We controlled absolutely the cost of production, but then our product was subject to railway freight charges which, like everything else, were constantly rising. Merchants' distributive costs, too, were not inconsiderable, bearing in mind that they had to break bulk and sell in small lots to millions of individual homes.

We set to work in co-operation with British Railways to see if some methods could not be adopted to ease costs and improve efficiency. At that time Dick (now Lord) Beeching was Chairman of British Railways and was deep in the job of reorganising his industry. His plan fitted perfectly with our own, because it called for the closing of hundreds of stations and the concentration of traffic at others. Coal from the mines then went in odd waggons to merchants. These waggons were treated as coal warehouses until they were emptied and then returned. Waggon use under this system was minimal and, as trains were made up at collieries, they had to go through further marshallings later in order to be broken up for their various destinations. Full train-loads of coal that could be quickly discharged and returned was clearly the answer. But few merchants could take whole train-loads and those who could would not be able to empty them and return the waggons quickly. The solution was the construction and development of centralised mechanised coal depots, each serving a large number of merchants. This would make it possible to work full train-loads from the collieries to the depots, eliminating the use of marshalling *en route*, and, by the use of automatic mechanical means, to unload the coal virtually on arrival, leaving the locomotive to pull out with the empties. The coal would then be stored ready to supply the merchants from a nearby depot and not from the colliery perhaps two hundred miles away. This minimised the risk of transport breakdowns in winter.

The work began with complete co-operation between the Coal Board and the railways: each had so much to gain from the achievement of the plan. Soon the engineering and technical problems were solved, and by 1963 the first of our mechanised depots was ready. It was at West Drayton in Middlesex, and was opened by Richard Wood in the presence of Dick Beeching and me in December 1963.

Announcing that £16 m. would have been spent by the end of the following year on the construction of similar depots, Beeching said that 80 depots were planned and 300 smaller ones (with rather less mechanisation) would also be brought into operation. To the railways this would mean that coal could be brought from the collieries in block trains and would avoid four or five marshalling operations *en route*. The West Drayton depot was serving an area which had previously relied on 23 small distribution points, using ten times the number of waggons. The benefit to the Coal Board would be that some of the financial savings would be passed back as a rebate on the freight charge, helping us to achieve our aim of stable prices for the housewife.

By 1970 the Board, the distributive trade and British Railways had in operation 60 fully mechanised depots with an annual total capacity of more than 5 m. tons. There had also been further progress in rationalising non-mechanised depots, with the result that their numbers had been reduced to only 385. When Beeching's original scheme was produced in 1962, there were about 5,000 stations handling coal traffic: by 1970 no fewer than 4,300 of them had been closed. Virtually all the suitable tonnage is now handled at concentration depots and the rest goes through colliery land-sales, by sea, or through railway stations which are to remain open.

Obviously, this gigantic reorganisation didn't come about completely smoothly; there were numerous complaints from the coal merchants about the added inconvenience they suffered as a result of the change. But by discussion and persuasion the task was accomplished. If this work had not been done, the cost of domestic fuel would be much higher than it is today. The Beeching era on railways was not totally liked, but there is no doubt that it marked a complete change in operations connected with coal movement. Coal is of course railways' biggest and, incidentally, most profitable customer, and the movement of millions of tons of coal each year represents a good deal of operational efficiency on the part of the railways and efficient organisation of the colliery sidings by our own people.

There followed a major co-operative effort for coal movement to the power stations by the Central Electricity Generating Board, British Railways and ourselves. This was the introduction of the 'merry-go-round' train to run between the pits and the big power stations. The system is based on permanently coupled trains which can load in minutes from overhead bunkers at the colliery and also unload speedily from bottom doors in the waggons into hoppers at the power stations. The trains make a number of round trips a day between pits and power stations as far as fifty or seventy miles apart, and need never stop moving—hence the 'merry-go-round' name. This was recognised at the time as a big breakthrough in freight haulage as far as this country was concerned and, of course, made possible tremendous reductions in the number of waggons that the railways had to handle.

The fullest benefit has not yet been secured from the capital investment in waggons and installation owing to the late commissioning of the big new 2,000 MW power stations, but the decisions were sound, were taken at the right time and certainly helped to keep down the cost of coal under the power station boilers. Obviously, this was of distinct help to the coal industry. Power stations were brought into service or taken out of service during the twenty-four hours of the day in accordance with demand, so that the expensively fuelled stations came in last, went out first and used less fuel than those that were cheaper to run. This is called merit rating. Fuel costs also had a big impact, of course, upon the choice of fuel for new stations. We were relying absolutely upon an increase in requirements from the power stations to keep our pits open, or at least to prevent them closing more quickly. Always at the back of our minds was the nagging fear of redundancies at the mines. We therefore had to fight oil and nuclear power—our rivals for this business—with every weapon that we could command. And we certainly did.

The great struggle with nuclear power comes later in my story. In the case of oil, it was not only the price difference between it and coal that mattered, but the capital required for a new oil station was less than with coal because there was no need to build expensive handling plant. Oil also gave the power stations greater independence from the manpower-intensive railway and coal industries. We therefore had no one 'rooting' for us but ourselves.

A 2,000 MW station burns 5 m. tons of coal a year, and that means the employment of about 10,000 men. More and more the CEGB were finding it difficult in the 1960s and onwards to get planning permission for new stations, and I weighed in to help them where a coal-fired station was involved. It was a mixture of propaganda and politics. It didn't always succeed but we weren't without our victories. One example was the Ratcliffe station near Nottingham. The planning authorities accepted the proposal, but objectors were quite numerous. A station at Holme Pierrepont, also near Nottingham, had been rejected in the early part of 1962.

On a visit to the area that summer I therefore made a speech showing the economic importance of the Ratcliffe station to the nearby pits. I said that the jobs of 10,000 men would be put into jeopardy without the station but with it there could be guaranteed stable employment in producing the coal for perhaps twenty years. Coupled with this, the rapidly growing demand for electricity could never be satisfied unless new stations were commissioned in time. The Ratcliffe station was built in the end and is now one of the most efficient in the whole country. The speech was well received and got considerable coverage in the local Press. The supporters of the scheme got new ammunition and the opponents were less vocal because of the human problems I had expressed. Nevertheless, in their later enthusiasm for nuclear power, the CEGB have not been very ready to acknowledge that stations like this and others recently built on the coalfield are still competitive with nuclear stations on cost, both of construction and in operation.

Our political and commercial campaign to win more of the all-important power station business also paid off in the autumn of 1963 when Richard Wood, the most kindly and considerate of my ten Ministers of Power, on one day guaranteed the future jobs of thousands of miners in Yorkshire and the East Midlands by giving consent to the construction of two coal-fired power stations, one at Cottam, on the River Trent, and one at Fiddlers Ferry, near Widnes in Lancashire.

This decision gave great heart to the industry in the East Midlands and Yorkshire from which the coal for these two stations would come, and at the same time emphasised in no mean manner the co-relation of these big capital investment decisions with employment and—as far as coal was concerned—the future economic prosperity of whole communities. Dick Wood always saw this relationship and understood; most of his successors never did. The CEGB were naturally only concerned with their own commercial judgments and at that time (the position changed drastically later) oil had a big price advantage. The cost of maintaining unemployed mine workers and their families and the large cost of encouraging new industries to establish themselves in the area were problems for someone else. Unemployment among miners has been created because of the false assumption of the electricity industry that oil would always be cheap and easy to get.

That the total cost to the nation was never calculated was undoubtedly the fault of the civil servants at the top. The statute setting up the Ministry laid it down that one of its duties was to co-ordinate the work of the fuel industry. In this it failed lamentably, and, as the different industries were managed individually on a commercial basis, the failure cost the nation millions of pounds in abortive capital investment. In the chapter on fuel policy I deal with this in more detail.

This was a period of great activity: we were moving up to the 1964 election and, what with the political undertones that this created, and the increasing pressure from the oil companies and business slipping away, no holds were barred. We turned our attention to the local authorities and the hospital boards, who collectively were big consumers of coal and naturally susceptible to the blandishments of oil.

One thing that particularly annoyed me was that many local authorities, whose prosperity had been built up on mining, never even considered coal when choosing the fuel for their new buildings. I paid a visit to Sheffield and launched a vigorous and well-publicised attack on the policy of the City Council. I pointed out that Sheffield drew benefit from the wages of the 100,000 miners in the area, but that not one school built there in the last few years was heated by coal. After opening the new boilerhouse at a nearby infirmary where coal was used, I went along to see the Lord Mayor and presented him with a brief prepared by my local marketing officials detailing, point by point, the city's neglect of coal. I made it clear that I wasn't demanding that Sheffield *must* use coal; all I wanted was to be allowed to bid on fair terms for the business that was going.

We widened our attack on other authorities and got the co-operation of people in the industry who served on local authorities to press our case, using briefs that we supplied to them. My deliberately dramatic intervention in the Sheffield case was effective and we were allowed a fair chance of all subsequent new business, some of which came our way. This naturally had quite an impact on other local authorities, and our marketeers throughout the country were able to persuade many into giving coal a fair chance. We knew that whatever we had to say about the social consequences, local authorities and others were, quite naturally, influenced by price. Whilst the oil companies could play the top tunes on prices by discounting, we were unable to do so because of the non-discriminatory clause in our statute. If they really wanted the business, they just bought it by giving lavish rebates.

It was essential therefore for us to ensure that coal was burned efficiently and that the greatest amount of heat was extracted from a given quantity. For industrial consumers we quickly developed and expanded a technical and advisory service. Qualified fuel technologists were available free to industry, and at one time our service was making about 15,000 reports a year. The staff assessed plant performance and made recommendations on the kind of fuel that would best suit the equipment, the method of operation, and the installation of new plant. Often the result of their recommendations was to reduce the amount of coal used by a particular customer but, of course, this was all to the good. If we could reduce the costs of our customers, the chance of keeping them on coal was much better.

In the massive sales promotion campaign we did not, of course, neglect these industrial markets. We adopted in our advertising and other promotional materials the slogan, 'Progressive industry is going forward on coal.' We showed that, contrary to popular belief, coal could be burned smokelessly if the boiler equipment was up to date. I constantly plugged: 'It is not coal that is obsolete. It is the out-of-date equipment in which it is still frequently burned which gives rise to the accusation that it is old-fashioned.'

In all this we were, as I have said, immensely helped by our own employees. A very large number were on local authorities and hospital boards. In our trade union members we had the largest number of unpaid salesmen that any company in the country possessed. Once they were alerted and briefed they went to work with enthusiasm.

The new headquarters of the NUM, then being built in Euston Road, were quickly converted from oil to coal. This had been a supreme example of the choice of fuel to be used for heating never being questioned at the time of approving the architects' plans. Some miners' clubs (a small number, it is true) were found to have been converted to oil, and the local outcry that these revelations produced got them put back to coal.

At an early meeting of the Coal Industry National Consultative Council, I reminded the mining unions (all of whom are represented on it) that it was no use passively regretting the loss of outlets for coal whilst traditional NUM strongholds appeared to favour oil. The message that the coal industry had just as much right to use pressure groups as had the oil industry came as something new. It certainly came home to me personally when the consequences of our new policy were felt in my old constituency of Blyth in Northumberland. We switched our own buying, wherever we could, from manufacturers that had gone off coal and in one case to my knowledge took the colliery canteens' milk order away from the local Co-op who had converted their dairy to oil. My reply to the protest that came was simple:

'It cannot possibly have escaped your notice that if everybody followed your example and turned to oil, there would be no miners left and that this change would very seriously affect your 22,000 members. . . . On the specific issue of our milk supplies, the income we derive from our coal sales must be spent with coal consumers.'

Sam Bullough, the Yorkshire NUM President, joined in the campaign and led his members with great gusto. In a highly publicised protest he refused to drink in his local pub after it had succumbed to a bit of rebate hustling on the part of an oil company.

Our marketing people certainly had their tails up by now and never missed a chance. They looked ahead, studying the market, and sensibly looked abroad, where the reconstruction of bombed cities had produced a number of district heating schemes. The idea of district heating itself was not new; it had been examined before the war but nothing very much emerged. The Coal Board promoted the use of district heating in this country—that is, providing central boiler plants which pipe heat to a wide variety of buildings (houses, shops, factories, hotels, offices). Customers are charged for the heat that they use. About ten years ago this form of heating received a further impetus with the introduction of schemes to burn refuse along with the coal. In 1966 the Coal Board launched an organisation capable of designing these heating projects and servicing them, called Associated Heat Services. The other partners in this enterprise are Compagnie Générale de Chauffe (who have had more than forty years' experience with district heating in France), and William Cory & Son Limited, whose knowledge of fuel technology and marketing is extensive.

There are now about seventy coal-fired district heating schemes in operation all over the country and others are being designed and discussed. The other fuel interests have followed our lead, but coal-fired schemes are easily in the majority. The biggest district heating scheme in Europe is the one that was designed in co-operation with the City of Nottingham. This boiler plant burns both refuse and coal and has an output from the one plant sufficient to heat a town of 40,000 inhabitants.

Associated Heat Services have also provided another valuable service in the battle for sales. Provision of heat, especially in larger buildings, is a specialist job which the owners or company management are glad to delegate to competent experts. AHS therefore operate a contract heating service, accepting all the responsibility for operating, maintaining and even replacing boiler plant. They guarantee the required heat load at all times and thus relieve busy management of an extra responsibility. This business continues to grow, and AHS have contracts to supply and run about 170 schemes, including blocks of flats, offices, warehouses and a big variety of factories and industrial plants.

We were affected in our battle for business by bad forward estimates of their own demands by our customers. The gas industry, for example, in 1958 had indicated that their requirement of coal for gas making in 1965 would be 27 m. tons and a possible 30 m. tons by 1970. In the event the gas industry actually took 3·4 m. tons in the year ending March 1971 and by 1975 will virtually have ceased to take any coal at all. I am not quarrelling with the switch from coal to oil for the purpose of making cheaper gas nor with the use of natural gas from the North Sea. But the gas industry walked away from coal without compensation of any kind. If the Coal Board had been a private company acting completely commercially, we would never have invested such vast sums without a contract. The termination of the contract would have brought in its train proper compensation for abortive capital invested. Thus the true cost of the fresh source of supply would have been revealed. Instead of that, the Coal Board was to be left not only with the capital investment, but also with the human investment, and if one remembered that about two thousand men are needed for every million tons of annual output of coal, this human investment, together with their families, was an immense one.

An alternative market therefore for the coal no longer wanted by the gas industry had somehow to be found. We had spent £4 million at Westoe Colliery alone (a Durham coastal pit), for gas coal no longer required. In total the investment in those big Durham coastal pits up to 1960 had exceeded £16 million—mostly for the benefit of the gas industry.

In 1964 it was decided to build a coal-fired station at Drax in Yorkshire. The original intention was to construct a 3,000 MW station (which would have made it the biggest coal-fired unit in Europe) and to have it in commission in 1969. This certainly was a prize for coal—a monster needing about $7\frac{1}{2}$ m. tons of coal a year at its peak. However the plans were changed and it was later decided to build a 4,000 MW station in two halves, the first to be commissioned in 1971 (two years later than the original plan) and the second, another 2,000 MW, at a later, unspecified date. This would raise the total annual consumption of coal to about 9 m. tons.

Naturally this was a development of tremendous importance to the Board and the Yorkshire miners, and, as each year went by and the second half of the station did not appear in the published plans of the CEGB, everybody got very apprehensive. The election of 1970 was looming up. The miners, Labour's most loyal supporters, were in a bitter mood. So far, the only impact of six years of Labour rule on the mining industry was the closure of pits, the diminution of employment and anxiety for the future.

By that time Harold Lever had become the Minister responsible. He was head and shoulders above all the other Labour Ministers who occupied that office, both in intellectual capacity and in speedy decision-making. He announced that he had approved the building of a nuclear power station at Heysham in Lancashire, a second nuclear power station at Sizewell in Essex, an oil-fired station at the Isle of Grain, and the second half of Drax. Roy Mason, the former Minister, claimed that the decision on Drax had been arrived at before he left the Ministry. All I can say is that it's strange that he didn't announce the decision at the time. After all, he was a miners' MP, and his stock with the miners was pretty low. It would have been a great help to the industry to have had the decision announced at the time he says it was made.

Although the construction of seven power stations had been approved since 1964, this was the only one to be coal-fired. But for the pressures that had been exerted, I doubt whether even that would have been agreed. There was no scrap of evidence at any time that the CEGB wished to remain with coal: indeed, all the comparisons with nuclear power were weighted by them against the use of coal. These figures ultimately proved widely out, but by the time this was admitted it was too late. The damage had been done and the pits closed.

A big outlet which it was inevitable we should ultimately lose was the gas market. Whilst a fair amount of research was carried out by the Gas Council in an endeavour to produce gas from coal much more cheaply than by the usual coking methods, nothing came of it and it was abandoned. Liquid methane began to be imported and then oil distillates as the feedstock for gas. To abandon the coal research was a great mistake in my view, as the processes made the whole of the gas industry dependent on imports. It was a bad decision both because of its impact on the balance of payments and from the point of view of security of supply. Despite the fact that in the USA they have indigenous oil and gas, their work on the gasification of coal has not ceased and it will not be long, with the rapid increase in oil prices, before an economic process is found.

The discovery of North Sea gas partly brought Britain's gas supplies back into the indigenous category, but this was more by good luck than good management, and was certainly not planned. (I say 'partly' because most of the firms involved are foreign-owned, so a large share of the profits flow out of the country.)

It was early in 1961 that the Gas Council submitted to Richard Wood, then Minister of Power, a scheme for importing natural gas from the Sahara. Quite clearly, this was a challenge that we had to meet, and we made a counter proposal that gas produced by the Lurgi process of total gasification of coal, then being introduced into this country, could well turn out to be as cheap as the imported methane.

In the end, the Government approved the Gas Council's scheme, but both the Saharan gas and that produced by Lurgi in the event quickly became commercially unattractive, first because of the introduction of plant using oil as its basic raw material, and then later because of North Sea gas. One Lurgi plant based on our opencast site at Westfield in Fife continues to operate.

At no time were the relations between the Coal Board and the gas and electricity industries on a proper commercial basis— a position that I inherited and found impossible to change. The original Act constituting the Coal Board laid upon it a legal obligation of 'making supplies of coal available, of such qualities and sizes, in such quantities and at such prices, as may seem to them best calculated to further the public interest in all respects, including the avoidance of any undue or unreasonable preference or advantage'. As a result, the mining industry could be treated as a warehouse by gas and electricity. We were expected to have coal ready for them; but if for some reason they didn't need it, we were left holding the stocks. Had supplies of coal to these two large customers been on a contractual basis, they would have had to take the coal or pay for it. As it was, they could happily leave the unpaid bills with the Coal Board. And, of course, if the nationalised industries had really been managed in the wider national interest, no one of them would have been able to inflict such serious damage on another. Unfortunately, Governments of both parties were eager to snatch at short-term advantages without thought for the future, and their civil service advisers were never in post long enough to do any real studies in depth. Fundamentally, this is the reason why none of the forward planning of fuel done by any Government since the war has been anywhere near accurate. Talented amateurs were dealing with long-range problems that only the experienced professionals running the industries were capable of handling. In most cases the professionals were consulted, but the decisions were still made by others.

The first North Sea gas find in September 1965 was, naturally, a significant event for the coal industry as well as for the country as a whole. It came from the very first hole put down by British Petroleum. And, although it was by no means certain that the floor of the North Sea was going to provide a valuable addition to the country's energy resources, I had to consider the effect on coal. Our existing competitors, oil and nuclear power, were being joined by a third. This was a marvellous discovery which, properly used, could be of enormous advantage to Britain. In the event, in my view, it was improperly used and the day will come when a generation will regret the profligate manner in which this new natural wealth was consumed.

I came to the conclusion that the gas industry should not be permitted by default to be the only nationalised industry involved in the exploitation of North Sea gas. The story of the Coal Board's extremely successful activities here is told in Chapter Sixteen.

The finds in the North Sea added yet another problem to our

marketing difficulties, because there is no doubt that in the first flush of intoxication the gas industry experienced over the earliest North Sea finds some ridiculous forecasts were made. For example, early in June 1966 spokesmen for the Gas Council were prophesying that the cost of gas to the consumer might fall to between half and two-thirds of the prices then ruling once the production of natural gas was under way. When the Coal Board gave their evidence to the Select Committee considering North Sea gas, we suggested that the capital cost per therm could not be less than 1s. od. The Gas Council, although they had plenty of opportunities to do so, never contradicted our claim.

Over-optimistic promises of the kind that I have quoted were being freely made in the run-up to the 1967 Fuel Policy White Paper, and I wonder what is thought now by people who were influenced by them into installing gas central heating. Far from cutting the price of gas by a half, the Gas Council have in fact greatly increased it. Nor were the quotations I have chosen rash, ill-considered forecasts. Later in that month of June 1966 the Gas Council officially stated that, as a result of the North Sea finds, about twelve million domestic consumers in Britain were certain to have 6d. a therm knocked off the price in five to ten years. Five years have now gone by and the Gas Industry still have another five in which to justify those promises. To do so they will have to bring about a sensational reversal of present price trends. Sir Kenneth Hutchinson, then Deputy Chairman of the Gas Council, even went so far as to say that after the first reduction of 6d. per therm, it was 'very probable' that the price could be slashed by a shilling off the charges then being made.

Promises of this kind had a very big impact, not only on the public but upon the Minister of Power and his advisers. The public were also regaled with a massive advertising campaign which talked about 'high speed gas'. I enjoyed commenting at a press conference that the gas had been going at the same speed for years and to me it was merely 'an old flame tarted up with a mini-burner'. Another advertising slogan for gas was: 'Heat that obeys you.' A lot of people's experiences of conversion to North Sea gas have turned that into a grim joke. As far as Government and the then civil service advisers were concerned our difficulties were made all the greater because they seemed to accept uncritically the gas industry's estimates of cost, and we had to convince them that coal was still going to be an important supplier to the British energy market. And we also had to maintain the confidence of the people working in the industry and the people we wanted to recruit.

I had always doubted the ability of the gas industry to bring in the North Sea gas and to convert domestic and other appliances at the speed they forecast. It was obvious to anyone with a grain of commonsense that it was just not going to be possible to recruit and train people fast enough and in big enough numbers. When I spoke to the Scarborough annual conference of the NUM in July 1966 I demanded to know: 'Where are all the men with spanners to come from?' In fact, the early conversions were so unsuccessful and troublesome that the Gas Council at one time had to call a halt until they had been able to prepare more thoroughly for the gigantic job that they had given themselves with the enthusiastic backing of people like Richard Marsh, Labour Minister of Power, and his civil service advisers.

But apart from the practical difficulties, there was no reason why the gas industry should have set itself such a tremendous task. The gas so far found in the North Sea is certainly a very welcome addition to our resources but the reserves so far proved will support consumption at the planned rate for perhaps only twenty-five to thirty years. Why did we have to run down the coal industry at such a speed so that we could get our hands on an asset which in any event will have such a short life? Time and again I argued that matters like these should be tackled with the 'total sum' approach. On the one side of the books there would be the saving to the balance of payments made possible by the use of North Sea gas. But this had to be offset by the fact that most of the profits would become a charge on the balance of payments because the companies involved in the North Sea were mainly foreign-owned. The 'total sum' ought also to take account of the writing-off of collieries that had been specifically reconstructed at great expense, using public funds to supply coal that the gas industry would no longer need. Many thousands of miners were made redundant because the coal industry lost this important market, and they had to be paid redundancy and other benefits by the Exchequer. I am sure that if all these factors had been weighed objectively one against the other the decision would have been to bring the gas ashore and to convert people's homes and factories at a much less breakneck speed. This would have also meant that the gas industry would have phased out over a greater number of years the enormous demand that it made for scarce capital that was needed for all sorts of other valuable purposes. Unless many more discoveries are made, this country will have a magnificent supply system converted to natural gas, but in twenty-five to thirty years' time there will be no North Sea gas left to feed into it.

Liquified methane or gas made from imported oil distillates will then again be required to fill the gap. I am sorry for the generation that will have to foot the bill, both for the imported raw material and the new plants that will be required to process it. More millions of pounds will be needed and at that point in time the competition for oil supplies will be such that there is every likelihood of demand outstripping supply. This will create no problems that will affect my generation but, as I said at the time, the comparatively short-term approach to the nation's problems, ignoring those of the future, could hardly be regarded as the approach of statesmen but rather of politicians.

The natural gas from the North Sea should have been given a life of something over fifty years: by that time other sources of energy now being worked upon will be fully operational. The real crisis point for the nation's energy supplies will be reached long before then. It would have been possible to have eked out the natural gas with gas from coal using new processes. However the die is now cast; a return to coal in the years ahead is hardly practicable because by then we shall have lost our most valuable asset, the mining manpower.

However, despite the set-backs, the whole team in our marketing department throughout the country fought tirelessly to hold what we had got. Most salesmen are selling in a growth market, but our chaps had to work in a contracting one. It didn't daunt their spirits and one day there came an unexpected bonus to encourage them.

A project to establish two aluminium smelting plants in this country was announced by the Prime Minister, Harold Wilson, in a speech to the annual conference of the Labour Party at Scarborough in October 1967. The case was based on import savings and furthermore it was expected that the smelters would earn substantial sums of money on exports of products. Aluminium smelting, however, required huge amounts of cheap electricity.

The general intention was that the electricity supplies would come from nuclear power stations—giving an ideal load as the demand was 24 hours a day and 365 days in the year. Two such nuclear plants were conveniently sited near the proposed smelters. But the snag was that once the decision to build aluminium smelters in the United Kingdom had been made, one of the world's biggest producers, Alcan, made it clear that they too would want to build a smelter here. Much confusion and embarrassment arose as a result of this decision, and at one time two competitive smelters looked as though they were booked for Invergordon.

One day when I was discussing coal industry problems with Harold Wilson, he said two things—why didn't we have a bargain sale of our coal stocks and why not try for one of the smelters? This was manna from heaven. What the Prime Minister did not know was that we had already made approaches and that at official level we had been politely warned off. The administration were determined that the smelters should get their electricity from nuclear generators. But Harold's suggestion broke this barrier down and we went ahead for the business.

The supplying pits had to be in Northumberland because there we had good reserves and the future markets for their coal looked like being short-lived. The smelters needed to have good access to the sea for their raw material imports and Blyth Harbour was close at hand. Furthermore, we owned extensive areas of land adjacent to the pits which we could make available to Alcan for their smelter.

So far so good. But still they needed cheap electricity and there was no way in which they could persuade the CEGB to provide this for them. Alcan preferred to build their own power station. The project was important to the Coal Board, not only because the coastal pits would continue to live and ensure security of employment for a considerable number of miners. Many pits had been closed in the county, many more were bound to shut and the alternative employment to be provided by the smelter would be enormously valuable.

No one rated our chances very high in the choice that had to be made between Alcan building a nuclear power station or a coal-fired one, but Derek Ezra, at that time the Board Member for Marketing, took the job in hand.

We put in an offer with a special price based on the forecast costs of the Lynemouth Colliery. The immediate battle was for an important piece of business, but it had a much greater significance than that. If we could prove that coal was cheaper for an aluminium smelter in the North-East, this would enormously strengthen our claim that the power station to be built at Seaton Carew near Hartlepool should also be based on coal. We all recognised too that Alcan knew the business of electricity generation backwards. In fact they were the biggest generators of electricity in the world apart from the public utilities. Only hard commercial facts backed up by their confidence in the ability of ourselves to supply the fuel would count with them. The first public hint that coal was going to snatch one of the smelters from under the noses of the electricity industry and the nuclear power interests came when Alcan submitted proposals in which they stated a clear preference for coal, which, they claimed, would be 30 per cent cheaper than nuclear on capital cost.

Now the fat was really in the fire. An independent, authoritative assessment of the commercial attractions of the two fuels had come down on the side of coal. This threw open again the whole issue of the competitive relationship between the two fuels which the Government hoped had been settled in their Fuel Policy White Paper.

The fury of the electricity industry had to be seen to be believed. The Electricity Council publicly and bitterly attacked our plan to supply cheap coal to the smelter. Smarting under Alcan's rejection of their argument that nuclear power was cheaper than coal, they claimed that it was unfair that we should supply coal to the aluminium smelter more cheaply than to them. They conveniently, in this argument, overlooked the fact that we had also made an offer to supply cheap coal to their power station near Hartlepool under a similar contract.

But the electricity bosses had another reason why they had only themselves to blame over the Alcan deal. They had been given the chance to supply electricity to the smelter from their coal-fired power station at Blyth, but had turned it down because of the extreme rigidity of their commercial policies. And the Ministry of Power lacked the commonsense to use their powers to co-ordinate the activities of the electricity industry and the Coal Board. Their reluctance to treat these two bodies as industries that had been nationalised in the public interest never ceased to astound me.

Coal for the smelter will be supplied at the rate of more than 1 m. tons a year from a twin-pit complex which links Lynemouth Colliery and Ellington. The two pits are now producing together well over 2 m. tons of coal a year. The coal, carried on a conveyor belt about a mile long, passes into the power station literally direct from the coalface.

There was a disastrous underground fire at Lynemouth in 1966 which fortunately did not cause any injury or loss of life, but which did cause the pit to be partially shut down. A new drift mine (a drift is a slanting access, as distinct from the more usual vertical shaft) was made at a cost of £2 m. and the two collieries are now linked underground. Lynemouth's fight back from the fire was one of the industry's big technical successes of recent years.

So when we did pull off the unexpected and started to be criticised by the electricity industry, I publicly gave the Prime Minister the credit, knowing that he would not be displeased. I knew too that if he were known to be behind our initiative, this would shut up our opponents. And that's exactly what it did.

Exports, of course, were not forgotten in the quest for markets, and in the summer of 1963 we were successful in getting new business from customers on the Continent. I travelled all over Europe at this time with Derek Ezra and Charles Howard, who was specially responsible for export business and is now the Head of the International Department.

We had a good response and added some unlikely customers like Rumania to the traditional buyers of British coal in France, Holland, Belgium, Denmark and West Germany. We also opened some promising discussions with the Italians, which later led to a nice power station contract there, re-opening business on a large scale after a lapse of many years. I had been confident that if Britain's application to join the Common Market succeeded, we would have smartly stepped up our exports to 10 million tons a year. However, General de Gaulle made us fight for our business on less favourable terms.

What clearly emerged however from these activities was that our ports were quite inadequate for doing business on a large scale. Apart from Immingham Dock, where a cargo of nearly 18,000 tons was once loaded very slowly and laboriously—the biggest ever in Britain—all other ports were restricted to 10,000 tons, the average being 6,000 to 7,000 tons, and loading rates everywhere were very slow—only about 500 tons an hour, so that waggons were held up. We were going all out for the Continental power stations, which were mainly dual-fired and switched fuels according to price. The freight charge was a big ingredient in the final price and this meant that we had to be able to load bigger ships. We therefore decided to go ahead with a scheme near Immingham but outside the Dock itself. The idea was to create the biggest bulk handling terminal in Western Europe. We also had in mind the possibility that it might be used to unload big carriers of iron ore for the steel industry, thus serving a dual purpose and further reducing their costs. The beauty of the scheme was that for the first time in our long maritime history coal trains would not need to wait for ships and ships would not have to wait for coal trains. With British Railways and the British Transport Docks Board, who were helpful and co-operative, provisions were made for full coal trains to empty upon arrival, conveyors discharging the coal in the stocking area. The trains never cease moving throughout the whole operation of unloading, which takes no more than 25 minutes for a train of 1,000 tons.

Ships could then come without waiting for trains and be loaded by a bucket-wheel arrangement which filled on to conveyors and then into the ship's hold at 4,000 tons an hour. The port was designed, of course, to handle coastwise shipments, mainly to power stations on the Thames and in the south, as well as shipments overseas. The scheme was carefully prepared and was well-based financially, providing a good return on the investment.

Because of the size of the capital investment it had to be submitted to the Ministry for their approval. What the Department did with it, I don't know, but it was eighteen months before it was returned, approved without the alteration of a single nut and bolt. In that eighteen months, of course, costs had risen by reason of inflation, but we were lucky—I suppose that this was during the period of 'dynamic government'; if it hadn't been, who knows how long we should have waited? Ultimately the British Steel Corporation decided to join us with an extension costing £6 m. for the importation of iron ore for the Scunthorpe steel works. This undoubtedly put the seal on a valuable national asset.

From my first few weeks in office the steel industry, then privately owned, was anxious to import coking coal, and I was quickly faced with an application to the Board of Trade by the Steel Company of Wales for a licence for American imports. The company were not seeking to buy all their needs from America, but simply to substitute American coal for the dearest British coal they were then buying. Obviously, the National Union of Mineworkers, especially in South Wales, were stirred up by this application. Other steel companies were also regarding this as a test case. Pressure on the Government to agree was immense.

My argument was that the British coal industry could never be properly developed and capitalised if the big buyers of coal were allowed to play the market, buying British coal when it suited them, and imports when they were temporarily cheaper. For months I could never address a meeting or a press conference without being asked questions about the claim of the steel industry that it ought to be free to import. In a speech to the Coal Industry Society I goaded the steel industry about their own 10 per cent protective tariff and was able to show that only small parcels of American coal, even with the freight market depressed as it then was, could be brought in at lower prices than a tiny proportion of the British coal used by the steel industry. I warned the steel makers that if they wanted to be free to shop abroad for the cheapest coking coal in the short term, they might return when price trends become favourable again to the British coal to find that our pits had been closed and could no longer supply them. We had invested heavily in pits to produce coking coal specifically to supply the blast furnaces. In South Wales alone, for example, no less than £46 m. had been invested. In his reply, Sir Julian Pode, managing director of the Steel Company of Wales and a member of the British Iron and Steel Federation's Executive Committee, had to concede that British prices for coking coal were comparable with those charged in Germany, France, Holland and Belgium. His case was based on American coal being brought into Italy. It seemed to me a thin argument to claim that the British steel industry was handicapped in relation to its competitors because one of them could buy some coal below our prices.

After long and earnest consideration the Government did rule against the application, much to the fury of Pode of Wales, who seemed to assume that I was personally responsible for the setback that he suffered. He and his colleagues kept up their pressure for months afterwards, but the Government held firm. Had the decision gone the other way, we should have had to cope with the industrial unrest that imports would undoubtedly have caused, along with an extensive stocking problem. It would have also led to a war between us and the steel industry, as we should have demanded the elimination of the tariffs on the steel which we used and shopped around the world for spot lots. This would have done neither industry any good, and the Government decision prevented it.

Nevertheless it was a fact that coking coals were much dearer than other coals and could be said to be overpriced in comparison to general purpose coal. At the first opportunity we were resolved to ease the situation and, we hoped, release the pressure which never ceased for coking coal imports. The opportunity came in 1964, when we were able to make the first—and as it has turned out only—price reduction in the history of the nationalised coal industry. What we did was to reduce the price of coking coal charged to the steel industry by 2s. 6d. per ton (about $2\frac{3}{4}$ per cent). The announcement was made by the Chancellor of the Exchequer, Reginald Maudling, to take effect later in the year. He presented it as part of the drive to keep down costs.

The Conservative Government in my earlier Coal Board years had been very considerate to the mining industry—indeed, upon reflection, far more considerate than the subsequent Labour administration. Selwyn Lloyd had imposed a fuel oil tax of 2d. per gallon which helped coal to the extent of more than £1 a ton and the Government had also imposed a total ban on imports. The first very helpful decision lasted throughout my tenure of office and the ban on imports had to be broken only because of the shortage of coal brought about by the Labour Government's Fuel Policy.

We had to run very fast in order to keep up with the demand for coking coal, and as the Durham pits in the west were exhausted and closed, the stock of available low volatile coking coal diminished to an alarming degree. As we dug into the seams further east and under the North Sea, the quality changed, and looking ahead we could see that imports of low volatile coking coal would undoubtedly be necessary in order to make the million tons plus of foundry coke, of which we were the only suppliers. Scientific research came to the rescue. Leslie Grainger, the Board Member for Science, was a thorough-going, enterprising and very down-to-earth scientist. He had been carefully examining the coking coal situation in general in terms of supply for the years ahead. Supplies for the steel industry did not create undue alarm, but something had to be done urgently about foundry coke. This was vital if we were to avoid imports. No time was lost by Grainger and his team at our Coal Research Establishment and the British Coking Industry Research Association. Innumerable samplings and tests took place, and eureka! the answer was found in a special blend of low and medium volatile coals. The resulting foundry coke produced from the blend which contained only 10 per cent of the precious low volatile was confirmed by the iron-founders as being quite equal to the coke made entirely from the low volatile coking coal. Indeed some went further and said it was better. Many of them had been using this foundry coke for some time without knowing that it was made from a blend and had noted no difference.

Sometimes research can involve lots of time and large sums of money without much result, but this was certainly not one of those projects. The research cost about £50,000 and we went ahead with building a commercial blending plant at Lambton in Durham next to our coking plant there at a capital cost of £300,000.

The success attending this research had not only averted the import possibilities but had prevented the closure of coke ovens in Durham and saved 3,000 jobs. It also saved the Kent pits now employing about 3,500 people. The Kent coalfield, which originally consisted of four pits and now has three, supplied the power station at Richborough and industrial consumers on the Thames. From the point of view of profitability it was a poor proposition; in the twenty-one years since nationalisation it had lost an average of nearly £1 m. a year on an annual production of only about 1½ m. tons.

On purely economic grounds there was no case for keeping the pits open, but the social problems that closure would have created kept them going. New methods of mining were being tried to make the pits pay their way when the blending process was discussed; and, in a survey of coals suitable for blending, the Kent coal passed the test. Arrangements were then made to withdraw the coal from the uneconomic markets, and during 1971 plans were going ahead to erect a new washery and coal preparation plant which would give the Kent pits a new and economically viable lease of life.

But the work that Leslie Grainger directed did not stop at solving the coking coal problems. He was tireless in his search for new uses of coal, and it was this that led to another potentially far-reaching discovery. As a result of trying to produce a smokeless solid fuel from ordinary bituminous coal (work which had been started by Dr Jacob Bronowski of TV fame fifteen years before), the research staff had acquired a vast amount of knowledge about fluidisation of coal which was being put to use in a number of directions. In layman's language the process makes coal flow like a liquid instead of behaving like the solid it really is.

We were looking for more economical ways of using coal for power station boilers and this led to a discovery of great importance. Work went on at our research laboratories at Stoke Orchard near Cheltenham on fluidised combustion. This consisted of a boiler where the water tubes were immersed in a bed formed of ash from coal and this was kept at a great heat by combusting finely ground coal and allowing it to fall like golden rain upon the ash bed. The advantages of this system were that the capital costs of the construction of the boiler would be less, the heat transference would be improved and coal would not need to be expensively prepared to reduce its ash content before burning. Experiments also tended to show that water tubes lasted longer as there was less corrosion. But perhaps the most important factor was that the fluidised bed retained the sulphur from the coal in the ash, and thus the problem of sulphur dioxide going up the chimney and creating pollution was solved. In the light of the general attack on pollution both here and in America, this was a wonderful discovery indeed. I can't say that the CEGB were over-enthusiastic, though. They had tied their future to the nuclear power star and here was coal muscling in again.

Our problem was that we were the suppliers of coal and not the users. Unless therefore we could get unqualified support from a big user, we would become involved in a research programme of considerable size without an end purpose. The lukewarm approach of the CEGB decided me to look across the Atlantic for a partner there. By this time the outcry against pollution in the USA had reached such proportions that the Federal Government under President Johnson had taken firm steps to tackle pollution and dereliction with a massive programme. I had long discussions with Mr Stewart Udall, Secretary of the Interior, which led to very close working between our scientists and theirs.

The American Government had set up a Federal Agency called the National Air Control Pollution Administration, and in 1969 I signed an agreement in Washington with them to exchange technical information on this system. The next year I again visited Washington and this time returned with a contribution of a quarter of a million dollars towards the cost of the research. I suppose, in a way, we had conducted our own piece of diplomacy to achieve this, and the result was not too popular in political circles here. What would have been more to their liking would have been for us to put the whole thing through the Government machinery. I decided to do otherwise because I knew from bitter experience that once anything of moment got into the bureaucratic machinery of Government, the results would take two to three times as long to achieve and we should be involved in endless meetings of working parties and hours of consultation. The face-to-face discussion, US Agency with Coal Board, proved to be, as I thought, far more effective.

The next stage should be a fluidised boiler at the 20 MW power station at Grimethorpe in Yorkshire owned by the Coal Board. Unless the CEGB take a really active partnership in this, the ironic result may well be that this British scientific effort, the first of its kind in the world, may first be developed elsewhere than in this country.

The battle for markets meant that all the weapons had to be used, and, not least, the little-publicised work of the research scientists.

The upshot of all this effort has been to confound the Labour Government's Fuel Policy White Paper and to justify the attitude of the Coal Board in its constant attack upon it.

When I was pressing the Labour Government hard in the spring of 1965 to maintain the coal industry at a healthy size, one of the arguments I used has since turned out to be only too well justified.

I said in our paper *Coal News* that a small coal industry would mean handing over our energy needs to oil imported from abroad. I added: 'This would be strategic madness, as Suez showed in a few short days.' Cut-throat competition from oil then being experienced was due to a glut, but I said that I was convinced that oil prices in parts of Britain had been only temporarily forced down by independent oil companies anxious to get a foothold here. We all know to our cost that the days of cheap oil have gone forever.

When the upward movement in oil prices finally came, it was like the bursting of a dam. In February 1970, the CEGB were claiming that if they were allowed to convert their Tilbury 'B' Power Station on the Thames from coal to oil, they would be able to save £3 m. a year. In fact, this was one of the conversions that we successfully opposed. The CEGB would probably now admit to some gratitude to us because, less than twelve months later, oil prices had soared by more than 60 per cent. Admittedly, under the long-term contracts they have with the oil companies, it takes some time for the CEGB to feel the full impact of price increases, but eventually the escalation clauses bite.

The Tory Government again refused to remove the tax on fuel oil in the 1971 budget; even if they had, coal in many parts of the country would still have been able to compete on price with fuel oil.

So my ten years ended with coal's competitive position transformed. Both on price and security of supply it is now well able to stand up to oil and nuclear power. The battle for business goes on, but it will not be against the seemingly impossible odds of a few years ago. Coal has proved its staying power, and its value to the nation is now being tardily recognised. There's no friend like an old friend.

5

Getting the Coal

Marketing, research, industrial relations, the ancillary activities, recruitment, training and administration were all highly important and fascinating facets of the business, but none of these could be allowed to take one's eye off the ball. This was the actual production of coal at the mines. As with all other aspects of business life, the most up-to-date tools to do the job had to be produced—for the accountant, the computer; for the senior management, operational research and the executive aeroplane; for the mineworker, something better than picks and shovels. The worker is rightly proud to claim the credit for increased productivity, but unless management is prepared to invest substantial capital in new tools and more modern techniques, then the limit of productivity with the old tools is soon reached. No amount of personal effort after that can produce more.

I was fortunate that by the time I reached the Board the mechanisation programme was under way. It was proceeding however at a very slow rate. In 1960 only about 38 per cent of the output was power loaded (in South Wales it was only 24 per cent, but in East Midlands it was 62 per cent)—that is, coal cut by machine and automatically loaded on to an armoured conveyor. The two machines which were the basis of the productivity drive, the Anderton Shearer and the Trepanner, were produced about the middle 1950s, but productivity remained on a plateau for a number of years. Too long—as subsequent events proved.

It is difficult, looking back, to understand the reasons for this. Certainly there was the demand for large coal for the domestic market that had to be satisfied, and the use of the machines seriously increased the proportion of small. There was a reluctance to depart from the known traditional methods of 'hand-got' to the unknown machine techniques, which produced their own additional problems. Labour was not too difficult to come by, price pressures had not arisen and there was certainly no central direction of planning. It had, like Topsy, 'just growed'. Indeed, so far from reality was mechanisation that a modern pit, like Lea Hall, near Rugeley in Staffordshire, which began coaling in 1960/61, was actually planned in 1951 for eighteen hand-got faces and an annual output of 1·5 m. tons. Today it is planned for an output of 2 m. tons from only seven machine faces.

What had been missed, I thought, was that the labour content of the cost of producing a ton of coal was over 60 per cent. With annual wage increases it was quite clear that unless the introduction of machines was quickened and a substantial improvement in productivity ensured thereby, then coal was undoubtedly going to price itself right out of most markets.

I had considerable pressure put upon me by colleagues not to take the steps I intended for the speeding up of the mechanisation of the pits, and a good deal of emphasis was put upon the supply position of large coal, of which each year there was an acute shortage for a brief period at the peak of winter. Indeed, when we entered the winter of 1961/62, there was some anxiety as to whether we would have large coal to meet the demands of the domestic consumer in the next few months. But what did not seem to be appreciated was the fact that the railways in 1960 were taking nearly 9 m. tons of large coal and that, committed as they were to dieselisation and electrification, their order for large coal would soon disappear. And indeed it did. By 1970, our sales to railways were minimal, at 140,000 tons.

On the other side of the coin, however, I had a man after my own heart. He was the Director-General of Production, W. V. Sheppard, a first-class mining engineer and full of realistic enthusiasm for an increased tempo for mechanisation. As his advice tallied with my own inclinations, I ignored the antis and went for machine mining as fast as it was physically possible.

I then found another snag and that was the organisation itself. Directors-General, although the Board's advisers and responsible to the Board for the day-to-day management and administration of the Departments they headed, had no authority to instruct the lower echelons but only to 'guide and advise'. The Divisional Boards had a degree of authority, but the real power lay in the hands of the Area General Managers, though even their authority was diluted by the two levels of authority above them.

I had to deal with this very obvious weakness in the management, which really meant that power and authority was so diffused that the decisions taken in the Boardroom lost their sting and impetus almost immediately.

To change the administration drastically would take time. It had to. I was a new boy and I didn't want to be regarded just as a new broom. And my first feelers had met with too much resistance at high level. I needed to play myself in before tackling a drastic re-organisation.

Better to take first things first; mechanisation stood out like a sore thumb.

Bill Sheppard, who became Board Member for Production in 1967, was at one with me in these matters, as was his successor as Director-General, Norman Siddall, another splendid mining engineer who needed no encouragement to pursue the new techniques of mining. When Sheppard was appointed Deputy Chairman in 1971, Siddall was invited by the Minister to become Board Member for Production.

Shep and I toured the coalfields and it did not take long to establish his authority, not only to 'guide and advise' but also to make sure that the job was done. The responsibility for any failures due to wrong policy decisions then lay squarely on our shoulders, as it should. These were great days for me because I was getting involved in the technical developments of the new machines and the new methods of mining at the same time as everyone else. Shep was a good mentor and I learned an enormous amount about actual mining operations. It was as good as having a personal tutor at a technical college, and I added to this by reading a tremendous amount of technical literature and visiting our technical research establishments. Both he and Siddall, as we travelled together, took immense pains to provide technical answers to my searching and, I am afraid, all-too-frequent questions.

Mining engineers are a tough bunch; they have to be, to do their jobs. They carry very heavy statutory obligations and are required by law to acquire special qualifications which means a specified time in the pits with actual coalface experience. I know of no other profession, unless it is that of a master mariner, where the training is so strict and the qualifying period so long. No man may manage a pit unless he has gone through this practical work in addition to his technical education. The result is that the mining industry has a large team of extremely competent staff who, having worked with men in the process of coal-getting during their qualifying training, are also good at man management.

Our pressures, then, started here, with the men who were responsible for producing the coal. These were the people we must enthuse and imbue first with the new ideas. So we arranged one-week courses at our Staff College at Chalfont St Giles in Buckinghamshire, and those courses that I was not able to address Bill Sheppard covered. In the short space of nine months we had between us addressed every one of our 682 colliery managers, agent managers and deputy managers. Even more important, they had been given the chance to question us and discuss the new policy with us.

Naturally, the part of the pit we concentrated on first was the place the output came from—the coalface. This and the road ends leading to it were the most dangerous, as well as the most crammed, part of the pit.

They are also the dustiest, and in the attack on the lung disease to which miners are prone, it seemed to us important, apart from doing the maximum amount of dust suppression, to reduce the number of men who actually had to work on the face. This is how the world's first remotely operated coalface was born. A fair number of experiments were being carried out by our Mining Research Establishment on the use of electronics and remote control, but it was going all too slowly. I sensed that the engineers were on the verge of something quite revolutionary, so I called a meeting in my room for a discussion. Bill Sheppard and John Adcock, then the Chief Mechanisation Engineer, were among those present. As they told me about all the new things they were doing, I listened, fascinated. I had to have the technicalities spelled out to me, but I did not need anyone else to explain the possibilities that they were unfolding. I could see an end to many of the dangers that threatened men working on the coal, and a great lessening in their physical burdens. All this in addition to a great leap forward in productivity. It was nothing less than a vision that my colleagues were offering me. The result of that meeting was our decision to go for a coalface that would be operated by remote control.

The engineers wanted to describe it as the Automated Longwall

Face, knowing that it wouldn't be long before it became known as Alf's Face. I thought this smacked a bit too much of the cult of personality and got them to think again. Their next suggestion was to call it the Remotely Operated Longwall Face, and ROLF it became. As early as the summer of 1963 we had two collieries in the East Midlands coalfield where this system was in operation. The cutting and loading and the advancing of the roof supports were all controlled from a panel 60 yards away from the coalface. The advantages of this were enormous. Not only was scarce manpower saved, but, even more important, men were virtually unnecessary on the face itself. The work done at those two collieries, Ormonde and Newstead, paid tremendous dividends. The ROLF equipment was also installed in a thin seam at Woolley Colliery near Barnsley.

We never did in fact get the full ROLF system working properly, because it proved immensely difficult to devise a method of steering the machine remotely so that it stayed in the seam and did not cut up into the roof and down into the floor. A man had to go along with the machine for this purpose. Many attempts were made to solve this problem, including a device which scattered gamma rays, but all to no avail.

However the 'spin-off' has been immense and is yielding dividends in many modified applications of the techniques used to produce the ROLF system. A few weeks before I left the Board, I had a letter from John Adcock, recalling the early days in which he thanked me for my courage in letting them develop ROLF. Now, six years after that historic meeting in my office, it looks as if the gamma-ray device has been made to work perfectly and, under certain conditions, the Anderton Shearer can now be steered automatically within the coal seam.

Although mechanisation has vastly raised productivity in the industry, the full benefits of the machines have even now still not been realised. Method study had shown in 1964 that coalface machines were standing idle much longer than they were actually running. In fact there were only 153 minutes of actual coal-getting on each shift—just over $2\frac{1}{2}$ hours of the six or so that the men actually spent on the face. On the average, 86 minutes were being lost each shift for no good reason. Another 70 minutes were being lost through holdups and delays caused by the conditions, and 45 minutes on jobs like machine turn-round and pick changing. It took the men an average of 75 minutes travelling during each shift to get to and from the face and another 25 minutes had to be taken up by preparation and a meal break. We were all realistic enough to recognise that in an industry like mining it would never be possible to have the machine running for every minute that it was available. But obviously 153 minutes could be greatly improved.

We therefore undertook our first big exercise in communications. To begin with, we sent every man a booklet setting out these facts and also giving him some information about the financial situation of the industry and the trends affecting it. There were sections on absenteeism and also on the loss to the industry caused through unofficial stoppages. All this was set out attractively and clearly, using diagrams and clearly written text.

We showed that even an extra quarter-hour a day on machine running time would be worth no less than £55 m. a year. We also showed how this would help to make coal cheaper and more competitive, thus maintaining the size of the industry and providing the maximum job security, as well as better pay and conditions for every man in the industry.

The booklet was only the start of a campaign. We also used our paper *Coal News* over the years, and the NCB are still using it for this purpose because it is a long-term job to influence people's attitude to their work. But always, this message, and any others which we sought to project to the people of the industry, was expressed in their own language. The skills of journalists, film-makers, designers, cartoonists, photographers and illustrators were all involved in this.

We were always on the lookout for new ways of communicating information about the industry's progress and problems. The most recent device we adopted is what has become known as 'the little black box'. This is a small cassette player, and we equipped all 290 collieries with one. When we decided to record a message, 290 cassette recordings were made by a high-speed process and posted to the collieries. They were played over at meetings of the Colliery Consultative Committee, and were intended to lead into discussion at that particular pit. It would be useless, of course, to suggest that any formula we might present through the cassettes was a solution, for example, to a particular safety problem and would be suitable for application at every pit. There is great danger in appearing to assume the rule of 'Big Brother'. So we took care that the tapes provided only the background information on which a positive discussion could be conducted at each colliery, so that solutions might be found to suit the local conditions. Obviously, it is an impersonal means of communication but it is also a novel one. We avoided over-using it and it will probably have a fairly limited life before it is replaced by some other means. But it is one more medium for the communication of ideas and information.

It was by this means that opportunity was given to the miners at the pits to hear what I had to say to their delegates at the 1970 Union annual conference. They were able to hear the simple economics of the industry restated and my attack upon the militants and their unofficial strike methods, together with the catcalls and the interruptions. It was, I am told, a great success, with requests for relaying to a larger audience than those who heard it first.

The rank and file in any organisation work much better if they understand what its aims and problems are. More important still is the individual commitment to the individual job. In the mining industry, what comes into the till is determined primarily by the regularity with which those coal-cutting and loading machines move up and down the coalface every day. In the last analysis, the commitment of the man who drives that machine, and of his mates who ensure that other operations keep abreast of him, determine whether we are in business or not. Coming back again to first principles, it has always struck me that physical conditions underground are for us rather than against us. During my ten years with the National Coal Board, I visited hundreds of coalfaces. All of them have had one thing in common: it was much more agreeable to keep moving along that coalface than to sit on one's backside. I don't think that any mineworkers would disagree with me on that. Why then is there still a gap despite all the improvements we have won between actual machine-running time and the time that it is capable of running?

We can discount any notion that miners are naturally bloody-minded or temperamentally ill-disposed to the job they have to do. Men no longer go into pits because there is no other job for them. Most of the time we had our work cut out in virtually every coalfield, in the face of fierce competition for labour, to recruit and keep the men we needed.

The answer seemed to lie in the traditional emphasis in British industry generally on direction and supervision rather than involvement and motivation. Not that we can afford to be too long-haired when it comes to managing a pit: somebody has to lay down standards and ensure their observance, otherwise the pit blows up. But over the past decade it has struck me that throughout industry generally, changes in workers' attitudes, the desire to be involved to the maximum extent in decisions traditionally regarded as the prerogative of management, have overtaken changes in the attitude of management themselves.

Looking back over the whole field of industrial relations, I think it may well be that, in our efforts to harness the goodwill of management and workers, we gave insufficient attention to the vital supervisory level between. In terms of financial rewards alone, we had to wait until we had rationalised (through the National Power Loading Agreement) the operators' pay structure before we could build a supervisory pay structure on top of it. Up to then, many first-line supervisors were earning less than the men they were responsible for who were on piece-work. In many coalfields they continued to do so until the full pay structure had time to take effect. Money, of course, does not count for everything, but it is an important ingredient (to put it mildly) in securing goodwill. We are only now, after twenty-four years of nationalisation, in the right situation with the first-line supervisors in the mining industry. Peaceful revolutions take time; this one has been no exception.

Six months after I became Chairman I went to have a look at the activities of the off-shore boring tower which was proving the reserves of coal under the North Sea that were to be worked by the Durham coastal collieries. Durham is really two coalfields. The coking coal on the west is in thin seams down to 18 inches (imagine spending a working day under a roof only that high from the floor!) and, valuable as it is, has been rapidly worked out. On the coast however the seams are much more reasonable in thickness and some of the pits are among the most efficient in the country. If they were to have a continuing life, they needed to know what reserves they could gain access to under the sea.

On my first visit to the tower, I was shown a core of coal taken from the borehole 2,000 feet below the North Sea some miles off-shore between Sunderland and South Shields. I knew that the confidence of the Durham miners was suffering because of the many pit closures on the west of the coalfield, and I knew that I was going to meet some of them on my return to dry land. I therefore pocketed a piece of coal from the core about a cubic inch in size and, when I met the miners' representatives, held it aloft for them to see. I told them: 'On this the whole future of the Durham coalfield depends.' The picture was widely published in the newspapers in the area and televised and not only gave a vivid illustration of what the Board were doing to keep the coastal pits going, but gave a much-needed boost to morale. I freely admit I used all the media that lay to hand and lost no opportunities to communicate in the best possible way with the men and their families.

The enormous progress in the design and power of the machines used to cut and load the coal has removed most of the back-breaking toil. But there has been another piece of technical progress which has had an even more radical impact on the pits.

The usual method of supporting the roof at the coalface was by vertical hydraulic props supporting horizontal bars held by them against the roof. As the face advanced the props had to be withdrawn from their positions, heaved by hand to their new stations and the roof bars advanced. Then the props had to be reset. The conveyor had also to be stripped down, moved forward and re-assembled. All this required a good deal of physical effort by a fair number of men.

But the research going on constantly between the Board and the mining machinery manufacturers had a big reward with the invention of hydraulic chocks which were self-advancing. These chocks are made up of groups of hydraulic legs—three, four, five or six in a cluster. As the coalface moves forward, a man turns one handle which first causes the vertical legs to retract. He then moves the handle to another position and the chock rides forward. Next, he turns the handle again and the vertical legs are anchored between the roof and floor in their new position. The conveyor is in hinged 'snaking' sections and no longer has to be stripped down before it can be moved. In fact, hydraulic rams shove it forward, section by section, at the same time as the supports are advanced.

This new method now covers nearly four-fifths of all longwall faces. It has reduced enormously the number of men on every coalface, so that it has contributed a big share to the productivity achievements. But, even more important, the new supports are enormously safer than the old ones. The men working at the face have virtually the whole of the roof covered by a vast, powerful, steel shield. Accidents used to be more numerous and more terrible at the face than in any other part of the mine. This is no longer so, thanks to the new self-advancing supports. They are more expensive than the old types, of course, but they pay for themselves over and over again in reduced accidents and higher output per man.

One big problem still to be mastered at the coalface is the driving of the roadways at both ends. For the cutting of the stable holes (as they are called) at each end of the face, to make room for the machine to be moved forward for its next cut, comparatively large numbers of men are still needed. And the speed at which these stables can be advanced determines the rate of advance of the coalface, which in turn decides how much coal is going to come up to ring that bell in the Coal Board's cash register. In some cases, the engineers have found ingenious ways of eliminating the stable holes completely, but this is by no means always possible, because of geological conditions. A number of machines have been developed by the Board's own engineers and by machinery manufacturers, but some are far too expensive for the job and most of the others are incapable of more than local application.

That the problem will shortly be solved, I am quite certain. And when it is this will make possible another big stride in productivity.

I found all these technical developments fascinating and it was possible as a layman sometimes to ask the right question that had an impact on the research. One could see sometimes with greater clarity, because one was standing back, that the technicians were beavering away and so obsessed with solving a particular problem that they had gone quietly ahead following a development that was inevitably going to aggravate an existing problem or even create a new one. The anxiety to solve the problem of mechanising the face ends produced a perfect example of this. Two of the Board's engineers had worked on a machine which they hoped would do the trick.

When I saw the prototype at work I was appalled by the amount of fine coal it was producing and asked for an analysis of the face output by size. This showed the amount of fine coal to be so great that, despite the fact that the immediate problem could be solved by the method, it would have produced too high a proportion of fine coal, making the face uneconomic. As we were not in business to make black flour, we turned aside to try other methods.

In a similar way, I once saw work going on to colour the coal so that it could be more easily recognised in its various household grades. Having been given the task to do, the scientists were doing famously and proudly showed me the result of various experiments. I asked for the programme to be stopped there and then: it only needed half an eye to see the cost that would come from the application of dye to thirty million tons of coal in five different colours. The enthusiasm of trying to meet the problem of distinguishing grades had completely over-shadowed the economics.

There is bound to be some uncertainty about coalmining. Obviously, a certain amount of information about the reserves of coal lying in a particular area can be gained by putting down boreholes from the surface. But between a pair of boreholes there may be what the geologists call faults, which throw the seam up towards the surface or down to a greater depth. Now that we are concentrating our operations on a smaller number of expensively equipped faces, management must try to eliminate as much uncertainty as possible. We developed a technique known as 'blocking-out' for this purpose. Under this system, roadways are driven right round the area that it is intended to work and even, in some cases, right through it. In this way much more information is obtained about the reserves to be worked.

There is also another method of proving the area before faces are brought into operation. The usual British practice with the longwall method of working is to advance faces into the coal, forming parallel roadways at either end of the coalface. This still involves going into unknown ground. Nowadays many more faces are worked on the retreat system. With this method the parallel roadways are first driven right out to the boundary of the area to be worked. The coal is then worked by a face linking the roadways at their further ends and by retreating back towards the main road and the pit bottom. Naturally, with the advancing method the roadways have to be driven and supported as long as the face lasts. Otherwise men, coal, air and materials cannot be got in and out. When working on the retreat, the roadways are abandoned as the face comes back towards the main road, since they are no longer needed.

One of the schemes where blocking-out has been used is at Longannet in Scotland. In the summer of 1963 all the news seemed to be good for the coal industry. One tremendous triumph was our success, after long months of very hard bargaining and pressure, to persuade the South of Scotland Electricity Board to build a giant station on the coalfield. This was the very imaginative Longannet scheme on the north shore of the Firth of Forth. The station would have a capacity of 2,400 MW, which was more than the then existing capacity of all the other stations operated by the SSEB. This meant, if we could get the business, that the jobs of 10,000 miners in the shrinking Scottish coalfield would be secure. Our planners produced the Longannet mining project, by which a number of collieries would feed their coal underground on to one massive roadway conveyor which came to the surface in the power station stocking yard to supply a large part of the station's needs.

The mining scheme is one of the most advanced and ambitious in Europe: apart from bringing the output of the collieries involved together underground, another revolutionary feature is the use of a computer to control all the loading facilities. This ensures that bunkers are not over-filled, that the main conveyor is not over-loaded and that the coals from the different faces in the scheme are blended to give the right mix.

But blocking-out with all its advantages still did not provide us with the fault pattern within the boundary roads and with the sophisticated system of mechanisation which a scheme of this kind involving millions of pounds demanded; we just had to have something more. Out of this came what is known as the Geosimplan.

Without being too technical, this is a method which combines elements of simulation and known facts and uses the principles of management games which are so popular in business schools these days. Knowledge of the faults (geological disturbances) in the surrounding region make it possible to produce the probable configuration of the area to be worked. This allowed our planners to estimate more accurately the best mix of high risk/high yield and low risk/low yield prospects for capital investment. International interest in the technique is already great.

The Longannet power station had a serious setback in 1971 but when it is fully commissioned it will use up to 5½ m. tons of coal a year, or the output of 10,000 men. It was certainly a piece of business well worth the effort and the time we spent on winning it, but it was the application of detailed and careful research and engineering that made it a physical possibility.

The fast drive to mechanisation however gave a great benefit elsewhere. It meant that less and less would pit ponies be necessary. Locomotives hauling more than 100 tons in a train load underground made the ponies redundant; and then came the growing use of trunk or main conveyor belts. Today there are 6,800 miles of roadways and 4,600 miles of conveyor belts underground. In 1960 there were still 7,750 ponies working in the pits; half-way through 1971 the number had dwindled to 620 and soon they will have all finally gone. Their departure from the pits is something that I'm sure causes mixed feelings for everybody. All miners who have ever worked in the pits with ponies have their favourite tales about their animal mates. I've heard yarns, for example, of ponies who've delivered their load of materials near to the coalface and then jumped on to moving conveyor belts so that they could ride back out. A tall story if ever there was one!

There have been societies formed for the protection of the pit pony, but I doubt if there are many working animals on the surface who are as well looked after, and receive as much affection, as the pit ponies do. Furthermore, their working conditions are checked by HM Horse Inspector of Mines. In contrast to the animal lovers, there are some people who become incensed about the concern shown for ponies which, years ago at any rate, ought to have been lavished on the miners. My attitude coincides with that of the Scottish miner whom I heard about several years ago. He gave a good 'roasting', as the Scots put it, to a youngster who he thought wasn't treating a pony kindly enough. The older man said: 'Remember laddie you don't have to work down here. The pony does.'

So, on the whole, I'm glad that it is now no longer necessary to take animals in to the pit to work. But I know the ponies will be missed. They were great pets of the men. Strangely enough they were good company too, and brought out the humanity in most of the men with whom they worked. It is a little-known fact, but the law allowed a man to work more hours than a pony. So the expression 'working like a horse' hasn't the same meaning underground as it has above ground.

I know that in most industries, when the boss is going to pay a visit to a factory or plant, a fair amount of special tidying up and preparation is done to impress him. It certainly happened at collieries when it was known that I was going to visit. Indeed, I have questioned managers about this. Most of them admitted it, but said that the work would have been done anyway; they had simply brought it forward because of my visit. I did not think that that was an unreasonable attitude.

In any case, with a pit, no amount of advanced whitewashing can guarantee a smooth-running demonstration of efficient and uninterrupted working. Geology just cannot be managed in advance—for the Chairman of the Board or for anybody else.

I remember, for example, on a visit to Bilston Glen Colliery in Scotland I was supposed to see a top-production face, but it ran into trouble just before I got there. The Shearer machine had cut up into the stone of the roof, having failed to follow an undulation in the coal seam. The cutter picks, instead of biting into reasonably soft coal, had been ruined by striking the hard rock above. When I arrived at the face the picks were being changed in preparation for a fresh start. Obviously I was interested to see how an unexpected snag of this sort was dealt with and I talked to the people in charge of the face and to the workmen themselves and discovered exactly what had happened and what they were doing to put it right.

In fact, of course, I learnt much more from that face than I would have done from one where the machine had travelled uneventfully backwards and forwards, cutting coal as easily as if it were slicing ham.

I found myself constantly chasing the difficulties. Of course, I was glad to hear all the success stories, but much more interested in discussing the mining problems.

I remember on one of my visits to the Nottinghamshire coal-field, I asked one of the mining engineers what his main technical problem was. This man said: 'What we have to avoid doing is letting the surface subside so much that the River Trent starts to flow backwards and empties the North Sea into people's back gardens.'

We went ahead with a determined concentration upon fewer but higher production faces. Unless we did this, we could never get the return on the big capital investment. To set off a coalface 200 yards long in 1971 with all the latest equipment of Shearer machine, armoured face conveyor and powered supports requires an investment of £175,000, so the coal has simply got to come and keep coming.

This policy of concentrating on a smaller number of high production faces has produced some results which would have caused eyebrows to be raised only ten or a dozen years ago. We have plenty of faces, for example, now producing more than 1,000 tons of coal a day. This was the output of a reasonable-sized colliery only a short time ago. A 1,000-tons-a-day face is in itself a sizeable business undertaking, representing a capital investment of perhaps £¼ million and having an annual turnover getting on for £1 million. Naturally, it has been necessary to adapt our management arrangements to suit these radical changes in the size and importance of our unit of production—the coal-face.

Coal coming off a face at this rate often also means a complete re-organisation of the haulage system. Mechanising the industry was not just a matter of designing face machinery and putting it to work. The fact of mechanisation affected the whole organisa-tion of the pit right up to, and including, the washing and preparation plants. It was a right-through task that put a great strain upon both management and men. It was essential to have a huge programme of training and this meant that, dealing with the coalfaces alone, about 10,000 officials and management staff were involved. We introduced what we called face reviews. Here at specially convened meetings the mining engineers responsible for the day-to-day management of the pit meet workmen, officials and craftsmen to discuss in detail the operation of the face; this means that everyone makes suggestions, the results forecast and actually achieved are reviewed, and so problems are solved. It became quite usual to hold one-day face conferences for every single man (and usually there are about seventy) who was to be involved in a new mechanised face. Very often a major conference of this kind reconsidered the working of a face already in production, especially if production was going down. We found that personal involvement to this high degree found not only the answer to the practical problems but aroused an enthusiastic support for all that we were doing.

It was indeed this participation by the workmen that enabled such incredibly improved results to come so quickly. To form the solid basis for the industry's future prosperity we selected fifty pits in 1967. The plan was to increase their combined output from 35 m. to 60 m. tons a year. The pits were chosen because of their potential, which meant not only their physical reserves, but also the quality and co-operativeness of the manpower. This was essential because we would need to pour a lot of capital into them.

In January of that year I set off to visit Wearmouth Colliery near Sunderland as the first of the fifty pits that I covered in a tour that finished fifteen months later, when I visited the last of them. Manpower was going to be a key factor in the success of the fifty-pit scheme, and at Wearmouth I said I was confident this problem would be solved once we finally got rid of picks and shovels and the hard, laborious and dangerous work, and gave the miners an adequate wage for the inconvenience of working underground.

Four years later, that is to say at the beginning of 1971, it looked as though we had got the manpower situation about right. For the first time in thirteen years our manpower rundown had been arrested, recruitment having improved (especially of the all-important youngsters that we needed for the future) and wastage having reduced.

In fact, during my ten years, I have seen manpower shortages and manpower excesses in the pits see-sawing with changes in the national economy. In all this, we have aimed to do two constant things: first of all to plug juvenile recruitment whether we needed adults or not, and second to keep a steady flow of men moving between contracting and expanding coalfields.

There has always been a tradition of mobility in the mining industry. You will find Welsh names on the payrolls of most

Durham pits, and hear Geordie, Yorkshire, and Lancashire accents in Kent mining pubs. In the old days, miners went where there were jobs, so the massive movement of men from Scotland and the North-East in the 1950s was merely continuing that tradition on a bigger scale.

By the early 1960s, though, the numbers of men moving had dropped substantially. Management in the traditionally 'exporting' coalfields were finding that they were short of men themselves, and the local unions were getting worried about the shift of voting power to the Midlands. To put it mildly, there was no longer any pressure from Scotland and the North-East for men to move south. One could see their point. On the other hand, if they were short of men in Durham, they were much shorter in Nottingham, where we were having to take skilled faceworkers away from their proper work to man unskilled jobs elsewhere. So we had to keep a steady flow moving.

I took the opportunity, on a visit to a Scottish pit with a very limited life, to ask to see the glossy booklets the Board had produced to encourage men to move south. Two hours later, the manager and his staff emerged covered in dust from the recesses of the pit stores and proudly produced one. Obviously we were up against a solid alliance of local management backed up by the local unions. We could send out national instructions until we were blue in the face. Nothing was going to change local attitudes. We needed some means of getting the message through to such men as would be willing to move (and we were not talking in terms of denuding the entire Northern coalfields!) without undermining the authority of their union or the dedication of their management.

And so we devised the Pick Your Pit Scheme. We built five mobile employment vans equipped to move into any closing colliery in Britain, establish telephone links with receiving pits, and fix up men and their families with jobs, houses and schools in the central coalfields. The employment officers also arranged visits for miners and their families to the new locations before clinching the deal. As often as not, we moved in-laws as well. By 1965, for the first time in many years, the movement of mineworkers between coalfields took an upward trend.

What was so different about the Pick Your Pit Scheme? As far as management in the exporting coalfields were concerned, the

Headquarters-controlled scheme was a much more orderly procedure than the old arrangements whereby importing Areas jostled with one another to poach trained miners whenever and wherever they could find them. At least local management now knew where they were in that they dealt with one central agency, and arrangements could always be made to defer transfers to suit the needs of the miner's colliery of origin. We dispelled union doubts by promising that we would only actively recruit at closing pits, and that we would also offer jobs at local long-life pits if any were available. For the transferee himself and his family (and this is what really mattered), the new arrangements were incomparably better. He could talk about the future in a relaxed way with people who were especially trained to advise him, and whose sole job was to see that the move went smoothly. We appointed in charge of the scheme the man who had been the most successful poacher under the old arrangements—Jess Whatmore, Area Manpower Officer in the Staffordshire Area. He had gone to all lengths short of kidnapping to keep the Staffordshire pits manned up in the early '60s. Under the Pick Your Pit Scheme he applied all his expertise to moving men (in a slightly less devious manner!) to whichever colliery throughout the whole of Yorkshire and the Midlands was most in need of them. A measure of the success of the scheme was the remarkably low wastage rate of the men we transferred—well below the average for the labour force as a whole.

All seemed set for steady progress in redeployment when the Government introduced the Redundancy Payments Act in 1965. The Act was supposed to make it easier for workers to move to new jobs. In fact, by 1968, it had reduced the flow of transferring miners to a trickle. Whatever the good intentions behind the Act may have been, the hard fact was that a sizeable redundancy payment was a much more attractive proposition for many people than continuity of employment. Miners at closing pits would not go near our employment vans for fear of being offered 'suitable alternative employment' and thereby losing their redundancy payment.

In fact we would never have tried to withhold a redundancy payment on the ground that we could offer jobs in another coal-field, but you simply could not get this over to the miners. Often we were put in the strange position of making a redundancy payment, keeping track of the recipient until he had spent it, and then offering him the transfer to another coalfield he would not even look at in the first place. The real moral is that Governments might find it useful to discuss, rather more fully than they do now, employment legislation with employers before they introduce it. In my opinion the Government would have spent less money to much better effect if they had concentrated on improving transfer allowances for workers moving to new jobs. This is precisely what we had to do in an effort to counteract the incentive which the Redundancy Payments Act offered redundant workers to stay where they were.

But we still found it very difficult to recruit and retain enough youngsters. In fact, the situation was so serious that we had to do something about it or run the risk of letting the industry bleed to death for shortage of up-and-coming workers. So our industrial relations people worked out a very imaginative scheme under which we could offer an apprenticeship to every youngster who came into the industry. We had always had good response to the excellent craft apprenticeship scheme that we had been running for some time, but it was less easy to attract youngsters to come in and become miners. The scheme involved the addition of an apprenticeship for mining boys, as well as for the electricians and fitters and other craftsmen.

In fact, although this scheme proved attractive for them, one of the things we did latterly was to make further changes. It had been found that youngsters of the type recruited into the mining apprenticeship scheme were less willing to face the three and a half years or so training that it involved than were the craft apprentices. In particular, they disliked having to spend time on training for all sorts of jobs that in the end they were unlikely to work at. They wanted to get as quickly as possible to the coalface, where the better money could be earned, and the changes we made permitted this and at the same time maintained the necessary standards to make it possible for the youngsters to work safely, as well as efficiently, at the coalface. What we did was to introduce two schemes for mining boys. The first caters for the lad who wants a planned career: when he finishes his apprenticeship he will be ready for training in line management or skilled specialist work. This has been supplemented by a shorter scheme for boys who want to become skilled operators in one of the many good jobs on the surface or underground. Reflecting the highly mechanised nature of the industry, both schemes include a period of basic engineering training. This was a new feature.

We were always looking for ways of bringing the various levels of our huge team closer together. For a number of years we had an annual business conference, usually held out of the holiday season in Scarborough. The most senior management in the industry attended. Having decided what the industry's first urgent priority would be in the coming year, we designed a two-day conference, starting on a Sunday evening, to discuss it thoroughly. These Scarborough conferences were followed up in each coalfield with a similar gathering for people at the next level of management. I remember that when our Welsh colleagues held their follow-up meeting at Porthcawl they said they were having 'a Porthcawl Scarborough'.

One of the most successful Scarborough conferences related to the more efficient use of manpower in the industry. We could see that wages were going to go on rising briskly and also that the industry, because it was contracting, was always going to find it difficult to recruit enough men. So at the 1963 conference we showed how much capital it was worth spending to save a man. We concentrated on the most laborious jobs and also those that were the most tedious. Pit managements put up their schemes to Areas, which after being carefully examined came to us at Hobart House. We allocated each year a capital sum for these schemes and made a country-wide selection giving, of course, first priority to those schemes with the biggest savings.

Manpower saving scheme projects came to be known as 'Scarborough Schemes'. The name has stuck to this day. Ask a colliery manager to tell you about his Scarborough Scheme and he will know exactly what you are talking about, and probably add that all he needs is the capital to get all of them done that he has already planned. As a matter of interest, today it is worth investing £5,000 to save a job.

From Kent to Scotland, the great team of people worked at the task of getting the coal with an unusual mandate, a very difficult one at that. Their task was to contract the total output whilst increasing productivity at a steadily rising rate. In ten years they succeeded in reducing total deep-mined output from 184 m. tons at a productivity of 305 tons per man-year to 133 m. tons at 463 tons per man-year. A productivity increase of nearly 52 per cent—not bad going! (In terms of output per manshift the increase was even greater.) Although it was a big team it was an efficient and enthusiastic one and the men co-operated magnificently.

6

Running the Industry

I had seen something of the organisation structure of the National Coal Board during my four years as Parliamentary Secretary at the Ministry of Fuel and Power. Being at the Ministry is a somewhat different thing from being in charge of the business (as I have since had to remind a few Ministers and Permanent Secretaries!), but even at that distance I became aware during visits to coalfield and pits of the remoteness of top management from what was actually going on in the coalfields.

At vesting date, the industry had been saddled with a five-tier line of command with supporting services at virtually every level except the pit. The line stretched from the National Coal Board through Divisional Boards (broadly corresponding to the major coalfields) to Areas, and thence through Groups to the pits. The end of the war released high-ranking officers from the armed forces who at least had experience of working within structures of that kind, and a number of them were appointed to senior jobs in the industries which the post-war Labour Government nationalised; mining was no exception. This was one of the mistakes made by Clem Attlee, who was the guiding light in this respect. I suspect that it helped to solve one of his major post-war problems—what to do with the senior ranks on demobilisation. Many of them, I have no doubt, were first-class service chiefs, but they were totally inexperienced, not only in the physical problem of mining, but in the industrial relations which early in nationalisation loomed large and the marketing issues that emerged later. The NUM were quick to realise this, and so were those mining engineers who, as Area General Managers in direct charge of anything between ten and fifty pits, were in the hot seat and were the people who really mattered. They were a determined lot (they had to be) and in many coalfields took little time in showing their contempt for some of the Chairmen of the Divisional Boards to whom they nominally reported. It was common knowledge, for example, that Yorkshire Area General Managers met together the day after each Divisional Board meeting to decide whether the Division's decisions were acceptable or not. One of their number would then convey their collective views to the Divisional Board. I was familiar enough as a Parliamentarian with the concept of a Leader of the Opposition, but I must say it struck me that was an odd way to run a business undertaking like the Yorkshire coalfield!

By the time I became Chairman of the Board, the formal organisation structure in many coalfields had ceased to bear much relationship to the way the industry was being run. Some Divisional Chairmen had a grip on their Areas and exercised real control over capital expenditure policy and return on investment. But others contented themselves with a Post Office role, exercising judgment only to the extent that they occasionally decided not to press their Areas to get on with any aspect of the National Board's policies with which they disagreed.

Quite soon after I had taken over the Chairmanship, the financial results in one coalfield had taken such a depressing turn that I asked for a detailed analysis by the Areas. This revealed that the adverse trends resulted almost entirely from excessive spending on materials (coalface supports and so on) in one Area. I arranged to see the Area General Manager, and was accompanied to his Area Headquarters by two senior Members of the Divisional Board. On the way there I asked whether they had themselves taken up these trends with the Area. One would have thought from the assurances I got that they had never been off the Area's doorstep. After we had arrived at the Area Office, the senior of the two Divisional Board Members swept open a door, apparently to usher me into the Area General Manager's office. We were promptly submerged in a cascade of mops and buckets. It was the cleaners' cupboard.

I subsequently discovered, in conversation with the General Manager, that he had never before been visited by a National or Divisional Board Member for a discussion on his Area's financial results.

This was of course very serious. Estimates had been made of financial requirements for the year, and one of the special items was materials, as it bulked very large in the expenditures of the Board. This was the main point discussed on a visit to another Area General Manager. The telephone had been in active use from Areas I had visited on the previous day, and he knew that some time in the morning, as we went through the accountability meeting, I should come to it. And indeed I did. This Area General Manager was one of the old brigade; in the old days, when he served the private owners, he would not have dared to spend an old-style penny over the odds without permission. But times were changed, the industry was nationalised and he had just gone on buying without any regard for his original estimates, so that his unauthorised spending was in the region of £1¼ million.

The scene was set. I sat opposite to him at his desk and on my left was the Divisional Chairman.

'Now', I said, 'you have overspent on materials to the tune of over a million pounds. Would you kindly tell me why?' This was really the big moment, the moment he was waiting for, the moment he had sat up half the night preparing for.

He slowly opened the top drawer in his desk and withdrew several sheets of closely written foolscap, in his own hand. He began to read, item by item, what he had bought and the amount it had cost. There was a slight pause after the first item, but I made no comment. He passed on to the second, then the third and so on. The pause between each item grew progressively less until he began to run one into the other, with only an occasional look up at my face.

I was gazing at him intently, with my best 'pot' face on. Not a sign of emotion—just deep interest as his voice droned on. He managed about two and a half sheets and by that time the confidence of the night before, when he was compiling it, began to wear thin. Instead of getting an involved discussion on each item in which his technical knowledge would have made mincemeat of me, I was able simply to listen in deadly silence. He had no way of knowing how I was receiving his news. So his voice faded away and, riffling the edges of the remaining foolscaps, he said rather pathetically, 'And there are all these.'

'Where did you get the money from to pay for them?' I asked. He looked at me astounded. 'Well', he faltered, 'I just have the invoices checked and authorised in the ordinary way and they go forward to the finance people for payment.' 'But at the beginning of the year you asked for authority to spend £1½ m. on materials, which was given. Now you have exceeded the amount by just over £1 m. Who authorised you to spend the additional amount?' 'No one. I needed the equipment and purchased it in the ordinary way.' 'I see,' I said, 'you thought that as an Area General Manager, you could dip your fingers into the till, and spend without compunction or reference to anyone in authority. Quite a change', I said, 'from the "good old days" when you dared not spend a penny without authority! Don't you realise that if every one of your forty-three fellow Area General Managers had behaved like you, instead of a £20 m. deficit, the Board would have been facing a deficit of about £70 m.? At the end of the day I have to carry the can for this industry and in the future, if you have a requirement to spend more than the authorised amounts, you must make a case showing the reason why the additional expenditure was unforeseen and why it is necessary. We will then tell you whether we have the money available. In the meantime please keep your fingers out of the till.'

I had recommended him to take more care in the future. Actually he had no future. At the earliest opportunity I dispensed with his service.

The story made the rounds and lost nothing in the telling. The Chairman had been taken on by one of the oldest and most experienced General Managers, and I had made it clear that I was going to exercise my authority as Chairman of the Board and that syndicalist rule by Area General Managers had come to an end. They certainly had power and authority, but only within limits set by the Board.

This incident however was just as much a condemnation of the lack of managerial oversight by the Divisional Board as anything else. But whether individual Divisional Board Members were up to their jobs or not, the National Board found itself confronted by a vast Divisional organisation between the policies which it formulated and the Areas which were supposed to carry them out.

The first National Board meetings which I attended were occupied by discussions of Divisional financial results and pious expressions of the Board's 'concern' or 'disturbance' at Divisions' shortcomings. These expressions were conveyed, in suitably polished bureaucratese, by the Secretary of the National Board to the Division concerned. These motions were rather like pulling the beer handles on the top of the pub counter without troubling to find out whether the pipes were connected to the barrels in the cellar.

All this implies no criticism of my predecessors. They presumably operated in the climate in which they found themselves, and it was very different from the one I had to work in. It is, too, much easier to see the weaknesses of an organisation when you come fresh to it, than it is when you grow up with it. The fundamental weakness of this one, it seemed to me, was that it prevented the central Board from agreeing feasible objectives directly with the Area General Manager, who in the last analysis controlled, and knew the potential of, the collieries. It also made a farce of accountability, because Divisional results masked a variety of achievement in their producing Areas.

Still, it was futile to imagine that we could move to a more flexible set-up in five minutes, and I was determined to do what needed to be done—at least in the early stages—with the Divisional organisation we had. In this I had some splendid support from those Divisional Chairmen who saw themselves not so much as presiding over a Divisional Committee meeting (the good Chairmen, in my experience, regarded that as the least important part of their job), as measuring and co-ordinating the work of Area General Managers. In my first year as Chairman I visited every Area General Manager in his Headquarters (usually with the Divisional Chairman, Bill Sheppard, who had been a highly successful Area General Manager himself, and Duncan Rutter, then my Principal Private Secretary) and got to know them all personally. Trend statements going before the National Board began to carry not just the faceless geographical location of the Area, but the name of the Area General Manager. The business became personalised. I think the Area General Managers enjoyed being recognised for the powerful men they were. I found very little resentment and a great deal of friendly help and co-operation. One or two of them tried the same feudal baron tactics on me which they had found effective in their dealings with Divisional Boards, and even then their departure was usually arranged amicably.

Gradually during the first year, we got over to every Area management team the fact that the customer did not owe us a living and that we must produce and sell competitively, or go out of business. In fairness it must be said that management in the coalfields found this message much more palatable than some of my colleagues on the National Coal Board. There was a vague feeling about Hobart House that the rough and tumble of commercial life was all right for the oil companies but ungentlemanly for us. When I came to the Board we were fighting to hold markets on inflexible price lists, whilst the oil companies were quietly (and quite sensibly from their point of view) rebating below their standard price wherever they might otherwise fail to win the business. This was an uneven battle and not one that I proposed to wage. The answer seemed to me to hold a central rebating fund with which Derek Ezra, then the Director-General of Marketing, could fight back on the oil company's terms. But I had a major battle convincing the Board that they must let the Marketing man deal flexibly with it. They wanted every rebated deal to come to the Board for approval. I had to convince them that you couldn't even sell cotton socks on that basis. Similarly, when the oil pipeline was spreading rapidly across the country in the early 1960s and threatening some of our most vulnerable markets, it seemed to me that we need not be in too much of a hurry to grant wayleaves through our substantial land holdings. I was beaten to it by a solemn injunction to Areas, which had already been sent by a senior Headquarters official, that it was no part of the Board's policy to obstruct rival fuels. It may not be now, but it certainly should have been then!

Clashes of this kind were symptoms of a difference of view with some of my Board colleagues on fundamental organisational issues. It is a commonplace of management education that the three main operations of business are production, selling, and research and development. In the mining industry, like most others, what you produce is determined to a very great extent by what capital you are prepared to invest and where you are to spend it.

I inherited a system at the Board's Headquarters whereby three committees—Finance, Research and Development, and General Purposes—virtually handled all the Board's detailed business. (Meetings of the full Board, then as now, took place at monthly intervals and provided a useful forum for exchanges with the part-time members.) The Finance Committee vetted and approved capital schemes coming up from Divisions to a total of about £25 million a year. The Research and Development Committee allocated expenditure which then totalled £4·5 m. or so a year. The General Purposes Committee, consisting of all full-time Board Members and presided over by the Board Chairman, dealt with *ad hoc* issues and endorsed the work of other committees. The Finance and R & D Committees were presided over by the Deputy Chairman (the Chairman of the Board and the Marketing Member were excluded from both, and the Scientific Member was excluded from the Finance Committee).

The Chairman of the Board and two of his Board colleagues were therefore effectively prevented from participating in most of the policy-making that mattered. It was argued that as the Chairman of the Board presided over the GP Committee, which ultimately ratified the Finance Committee's decisions, he was merely being spared the burden of the detailed vetting of capital schemes. To my mind, this showed a misunderstanding, not only of a central Board's capital expenditure function, but of the priorities with which the Chairman of any central Board should concern himself. In one breezy interchange with colleagues at a Board Meeting I indicated somewhat forcibly that I could not continue to tolerate a state within a state. It seemed to me that if schemes still technically suspect had emerged through the tiers of financial specialists and engineers at Area, Divisional and Head-quarters level, then we should get rid of the specialists who had vetted them and find some new ones. As it was, the Finance Committee spent its time as if it were playing a parlour game like 'spot the deliberate mistake'. When it triumphantly found one the matter was taken up with the Area (through the Division) and then the General Purposes Committee was invited to approve the scheme. All this performance succeeded in missing the fundamental question: in the light of the industry's future objectives, should £X million be spent on this scheme rather than on another similar scheme, or, indeed, should any money be spent on that type of scheme at all?

By the time the decisions of the Finance Committee came to the General Purposes Committee for endorsement, it was too late to go back to square one; the real decision had already been taken. When I made noises on those lines, some of my Board Members appeared surprised that I wanted to be involved in fundamental business decisions. The impression in some quarters at Hobart House had apparently been that I was going to be content to be a political front-man whilst other people ran the business. That was an erroneous impression, especially as I was far from satisfied with the way they were running it.

I put it to my colleagues, therefore, that whilst I did not expect to be personally involved in all matters of moment, I did expect to have a say, both in investment and in Research and Development policy. There was, to my knowledge, no enterprise anything approaching our size anywhere in the world which excluded its Chairman and Marketing Board Member from two such fundamental activities. Unfortunately my request was interpreted by some members of the Board as questioning their competence. It became evident that they were preparing to force the issue to a vote, and my practised political eye detected some fairly crude lobbying of part-time Board Members. Equally I was determined not to give way on something so fundamental, and I privately resolved that I would, if need be, ask the Minister to make changes in the Board. Inevitably the issue was forced to a vote. The Board was split right down the middle, and I might well have been forced to take matters up with the Minister but for a telegram of support from Frank Wilkinson, the Board Member for Marketing, who had read the Board papers whilst on holiday in the South of France and could see that I was likely to have a battle on my hands. Frank had no axe to grind (he was within a year or so of retirement), but he had an instinctive flair for business and had seen only too clearly that millions had been spent in the past giving access to coal which was likely to be totally uneconomic in the light of future market trends.

The demise of the Finance and the Research and Development Committees brought to a head the differences of view between what had come to be known as 'the Old Guard' on the Board and those who were prepared to take a new look at our problems and the way we should tackle them. As a House of Commons man, I had been used to winning or losing fights of that kind and then settling down amicably to work after the dust of battle had subsided, so I harboured no grievances on this issue. Nor, it is fair to say, did most of the Board Members who had opposed me. Arthur John (at that time Finance Member of the Board and now Finance Director of Unigate) was generous enough to concede later that he thought there was a real advantage in taking a corporate view on two of these most important functions. From now on, all Board Members began to deal with the big issues.

The plethora of committees was abolished and when I left

the Board had but two: the General Purposes Committee, comprising all the full-time members of the Board, meeting every week and the Board itself, meeting monthly. At these meetings Directors-General of Departments who were responsible for the implementation of Board policy attended as the Board's advisers. When we wanted a special task performed or a particular matter looked at in detail, we used to set up a small working party which, when its task was done and it had reported, ceased to exist.

This brought the whole of the Board into close contact with every major decision and gave enough time to examine policy, providing at the same time an accountability service to enable it to carry out properly its statutory duties.

There remained the much more difficult problem of the way the central Board's policies could best take effect at collieries and, equally important, the way in which Area and colliery management's vast practical experience could be harnessed so as to make a full contribution to the formulation of policies. We were all agreed that the fewer the levels between the National Board and the Colliery Manager the better. On the other hand, it would have been quite unrealistic to expect to run more than 600 pits directly from Hobart House. If we had too many levels of management not only would initiative be stultified, but the industry would be incurring needless overheads. In 1963 nearly 3,000 'white collar' staff were employed at National Headquarters, 6,800 at Divisions, 17,000 at Areas, and 4,300 at Groups —in total 31,000 at the four levels above the pit. These staff were costing the industry at that time about £23 m. a year. The intriguing question was whether they were putting back into the till, in terms of direction or service to the collieries, as much as, or more than, they took out. If they were not, then the initiative for change must come from the top; nobody could reasonably expect staff voluntarily to declare themselves redundant.

I therefore suggested to my colleagues on the Board that we should propound for ourselves a 'philosophy of management' setting out precisely what needed to be done to produce, sell our product, and stay in business. We used the Board's Staff College as a forum for these discussions, and as a preliminary the College's then Director of Studies, Peter Tregelles, spent a year studying business practice in large organisations, both here and in the USA. The discussions were far from being confined to Members of the

Board; Divisional, Area and colliery managers took part, quite uninhibitedly, because we were at that stage discussing organisational principles, not structure. Views were expressed in the knowledge that, whatever the outcome of our deliberations, my Board colleagues and I were not hatchet men who would put organisational theory before personal and individual considerations.

It became clear that there were three principal activities in producing saleable coal on a national basis: operating the collieries; designing and supervising colliery operations so that each made the maximum contribution to the needs of the industry as a whole, and took account of the best available technical expertise; and finally, farming out the Board's business to collieries, allocating resources, agreeing production and profit objectives with the producing units, and holding them to account for attaining them. This seems simple enough, but it should be remembered that the organisation of the industry at nationalisation was based on an amalgam of military 'line and staff' at the top, and colliery company practice (where status was often more important than function) below. After fifteen years of that, it was not easy to get to first principles. Once an organisation is in being, it tends to justify itself. As I looked around Hobart House, to take but one level of the organisation, I saw a variety of people with prestigious titles, all superficially articulate at meetings, all apparently making a substantial contribution to our business; yet frequently, when hard facts were needed for a Board decision, the basic information was not there precisely because the man who had it was not considered of sufficient 'status' to attend Board meetings. The expression 'It's not my side of the house' went into the history of the Board's organisation because of my own rage when senior representatives of a Department thought this was an acceptable explanation for their failure to take any interest in the Board's affairs outside their immediate bureaucratic niche.

I would occasionally ask to see departmental files on a matter in which I had taken a personal interest and would find that creative work at lower levels was all too frequently being stifled by a top-heavy departmental structure in which every man in the hierarchy felt obliged to make a contribution. The contribution, unfortunately, was mostly negative rather than positive.

Often unnecessary work arose because of the high standards of functional skills in the industry. A centralised Purchasing Department had done a first-class job using our vast purchasing power to bring suppliers' prices down, and in keeping stock-holding points and levels to the minimum. But one had continuously to watch out for signs of departmental interests becoming divorced from those of the business they were to serve. On more than one occasion, an over-zealous stores controller had cost us as much in a day's lost output through lack of machinery spares as he could have saved us in years of stock control. I heard many suggestions in my early days at the Board that, because we had so many tiers of engineering services, needlessly high professional engineering standards began to be asserted for their own sake, whereas what the business really needed was a much more flexible approach to a rugged, extractive industry.

This kind of problem occurs in any large organisation, whether publicly or privately owned, and the only answer to it, to my mind, is a periodic review to ensure that staff are on the payroll only because they are making a positive contribution to the business as a whole. I think that initiating reviews of this kind, as well as seeing that they are objectively and fairly conducted, is one of top management's most important jobs.

After a couple of years in the Chairman's seat, I had had time to take all this in, and it became apparent that the then-existing five-tier structure of management—collieries–Groups–Areas–Divisions–Headquarters—could no longer be justified. With a decreasing number of collieries, there was less need for the intervening units of management between them and Headquarters.

In fact I thought that they tended to act as barriers in the management structure. But I needed the views and advice of those who had had greater experience. The opportunity came in November 1963, when I was on a visit to the Nottinghamshire coalfield. I was spending the night as the guest of our Divisional Chairman, Wilfrid Miron. Also present were Collin Cowe, the then Secretary of the National Board and a former Divisional Board Member, and Duncan Rutter, who had been an Area Secretary. So between us we had had considerable experience at several levels of coal industry organisation. When Mrs Miron left us after dinner, we chatted in front of the lounge fire about the need for re-organisation changes. Although Miron was himself Chairman of the highly successful and profitable East Midlands Division, he expressed views about how a lot of management could be eliminated in its then existing form. This relaxed fireside discussion ultimately led to a paper which I put to the National Coal Board, as a result of which the Committee on Organisation was set up: 'To examine the organisation of the National Coal Board in the light of developments in the industry and its future requirements', as the terms of reference put it.

As the whole future of Divisional Boards was at stake, it was essential in my view that they should be strongly represented on the Enquiry Committee. I chose two Divisional Chairmen who I knew would take an objective view based upon their immense experience. My choice fell on Jim Anderton of North-Western Division and Wilfrid Miron of East Midlands. There was no better respected mining engineer than Jim Anderton. His name is known throughout the world because of the most widely used power loading machine bearing his name and of which he was the inventor. The Anderton Shearer can be found in use in pits wherever in the world coal is mined. If he had developed this privately he would have become a millionaire; as it is he preferred to accept more modest recognition through the Board's Awards Scheme for Inventions.

Wilfrid Miron was a completely different kind of man from Jim Anderton. He had great administrative ability and with it a friendly personality that made it possible for him to be quickly on good terms with people. A Welshman and a solicitor by professional qualification, he spent just under two immediate pre-war years with Shipley Collieries, Ltd, and came to the Coal Board right at the beginning as Secretary and Legal Adviser of the East Midlands Division. Later he became Deputy Chairman, then Chairman of that Division, and is now a Member of the National Board. This wide administrative experience, coupled with Jim Anderton's unchallenged knowledge of coal winning, seemed to me the most qualified combination necessary for the task.

But who should preside over the review body? This was the sort of job which, in my opinion, was important enough in any organisation to warrant the personal attention of a Chairman or his Deputy. On the other hand, where the outcome is likely to involve the careers and well-being of a wide range of employees in the enterprise, the top men could reasonably be expected to look objectively at recommendations made to them by a working party, keeping the employees' interests, no less than organisational considerations, in mind. I urged this latter view on the Board's Deputy Chairman, and appointed Harry Collins, the Board Member for Production, as Chairman of the review body.

Collins, a former Yorkshire Colliery Manager who helped to put the Ruhr coalfield back on its feet after the end of the war, was widely respected in the industry. He combined sound commonsense with a shrewd notion of human strengths and frailties. There was nothing of the organisation man in him. If any man could tackle the job fairly, and be seen to do just that, he could.

The other members of the Committee were: Sir Reginald Ayres, a part-time Member of the Board, who had retired from the job of Deputy Secretary of the Ministry of Power after a distinguished career in the Civil Service; Arthur John, another Welshman, who was the Board's Finance Member and had served the Board since its inception; and E. F. Schumacher, the Board's Economic Adviser—a man with a penetrating mind and the ability to make people think again about judgments they had always taken for granted.

Collin Cowe, the Board's Secretary, was a member and also performed the secretarial function. Here again was a man of great capacity, whom I knew first as the private secretary to the Board's first Chairman (Lord Hyndley) back in 1947. He was destined for high office in the Board and had begun successfully to cut his teeth as Managing Director of one of the ancillaries (Associated Heat Services) when a terrible motor accident of which he was the blameless victim made it necessary for him to seek a quieter life when his long recovery period was over. He too had experience as a Divisional Board Member, having for five years been the Member for Staff on the North-Eastern Board.

As a result of the Collins Committee's recommendations we eventually introduced a new form of organisation to take effect at the end of March 1967. This reduced the numbers of management levels to three—collieries, Areas and National Board—instead of the previous five, and the number of Areas by about half to seventeen.

This was perhaps the biggest management re-organisation ever undertaken in a British industry of any size, and I often think back to its beginnings in that fireside chat in the Mirons' house three and a half years earlier.

I chose an internal committee as the instrument rather than bring in outside consultants, in the belief that in an organisation of our size there must be enough experience and knowledge of the problem for us to be able to find our own remedies. The Collins Committee spent about a year preparing its report to the Board. In fact, this was a remarkably speedy job considering that the Committee visited every coalfield and talked to management at every level. As I hoped, the report showed that it would be possible to carry through a most thorough-going reshaping of the organisation, and make the reduction in the number of tiers from five to three. The indications were that the number of collieries would be down to between 300 and 350 by 1970 (in fact, the lower figure proved to be a slight over-estimate). And, since experience had shown that an Area could effectively control about twenty collieries, it was obviously going to be possible to reduce the number of Areas considerably. In fact we settled on seventeen, which gave a solution suited to the geographical location of most of the pits. The next question was whether National Head-quarters could directly manage as many as seventeen Areas. Here, there was no guidance in the Board's own experience, but modern management theory developed elsewhere suggested that this would be perfectly possible.

Obviously, such an enormous operation was going to take a fair amount of time to prepare properly, bearing in mind as we always did that the impact on many thousands of people was going to be serious. So we decided that the beginning of the new financial year in April 1967 would be the effective date. The main principles of the new organisation had been decided in the autumn of 1965. Because the staff were going to be so deeply involved, the first job was to communicate the decisions to the industry, and in particular to the non-industrial staff. (The miners themselves were not going to be greatly affected.) Two conferences for all the senior staff were held at an hotel near London Airport on succes-sive days in December 1965. About 250 people had the Board's plans thoroughly explained to them at these conferences and immediately afterwards the leaders of the unions, representing between them virtually all the industry's workpeople, were seen and given the same information on paper for the record, as well as orally. The senior staff who had attended the airport conferences went back to their own Divisions and Areas and passed on the information, using briefs provided for the purpose, to the less senior staff in their formations.

Within a week this huge communications job had been completed and all the non-industrial staff had been given a full account of what was intended. But of course, this was only the start of the process. Thousands of our staff wanted to know what was likely to happen to them personally. Although we couldn't give precise answers to everybody, we were able to give some indication to most. Early in January all the most senior people from Divisions and Areas were interviewed in London by Members of the Board and the Head of their own Department. Each man was told what the Board had in mind for him and the opportunity was provided for objections to be raised and difficulties to be talked over. Then in March and early April the next level of staff were interviewed, also in London. No fewer than eight hundred people had been given a personal statement about their own prospects by the middle of April, and names had been pencilled in for all the senior posts in the new Areas and in Headquarters' Departments. Whilst all this activity was going on, similar interviews were being given to the less senior staff in the coalfields. The next step was to weld the new Area staff into effective units. Some members of a particular team would be well known to the others, but some were complete strangers. We therefore brought all the Area top-line staffs to our Staff College at Chalfont St Giles with two main purposes in mind—to give them a chance to get to know each other, and to tell them how the new Areas were to operate.

Re-organisation was handled with great care in order to minimise any hardship to the staff, as was the reduction in miners' jobs. Natural wastage was one big factor. Whenever anybody left for another job, he was either not replaced, or he was replaced by someone already in the organisation whose job was becoming redundant. Special permission had to be sought to recruit from outside. Departments were given staffing objectives along with their other management objectives. Some people were retired early (at fifty-five or over in the case of men, and fifty for women). The Government's contribution towards the social costs of contracting the industry was used for this purpose, as well as for taking care of redundant mineworkers.

In the six and a half years before re-organisation, it was possible to save 6,700 non-industrial jobs. However, in only three years after this enormous scheme was implemented, the staff numbers were reduced by 7,700, or one-sixth of the total. Again one tries to think through the statistics to what they mean in human terms, and we tried to deal compassionately with every single one of the people involved.

When we finally got started on the huge job of informing people in the industry about our re-organisation plans, it meant that, for me, two years of preparing the industry for these changes had come to an end. All my colleagues and everyone else who had been involved in the discussions accepted the case for these very sweeping and fundamental changes. The tremendously hard work of translating the new organisation from lines on paper into reality had still to be done, of course, but the quiet campaign of persuasion was over. Naturally, not everyone was pleased with the effect the re-organisation had on him personally. But the need for it in the interest of the industry as a whole was accepted generally by all.

With the new organisation, we also introduced a new technique of management by objectives. We helped managers achieve their objectives by providing a large range of the latest management services—a computer network, an operational research service, and an experienced team of organisation and method specialists. All the Area Directors were mining engineers by training and experience, but it was essential that they should become primarily businessmen, because they were responsible for undertakings that employed an average of about 20,000 men, and had capital resources of about £50 million, with a turnover higher than that.

The system of management by objectives that we introduced is a comprehensive and sensible one. Area Directors are responsible to the Board for the achievement of annual objectives set out for them in terms of output, sales and profitability, after full discussion and consultation with them. Then the Area Director, with the advice of his Business Planning Team made up of his Heads of Departments, breaks the Area objective down to colliery objectives. Again, there is consultation with the colliery management. Then a programme of the action needed to achieve the objectives is adopted. These action programmes cover a period of eighteen months ahead and are brought up to date each quarter. Some of the earlier forecasts proved highly optimistic, and it took two or three years before most people began to make reasonably accurate estimates of their results.

I used to make a point of asking, on my visits to pits, to see the colliery action programme. Almost always, when the action programme was being treated seriously, the pit was well managed. But quite often it was apparent to me that the manager regarded the action programme as a bit of a nuisance and something worked out by people who were remote from the colliery to complicate his life for him. Naturally I made a fuss when I came across this attitude, which boiled down to a feeling that the holder of it was far superior to such methods.

So important did I consider the confidence and morale of the industry that, when we re-organised the Board in 1967, I made Public Relations, which had hitherto been responsible to the Secretary, a Department on its own. Furthermore, I made the Departmental Head directly responsible to me. In this way I was able personally to look after what I regarded as one of the most important aspects of my job as Chairman.

One of the first organisational changes that was made, long before the big re-organisation, was to set up the Coal Products Division to manage coke ovens and chemical plants. Previously these activities had been managed by the Divisions but, since the main pre-occupation of the Divisional Boards was to get coal efficiently and profitably, the specialist problems of the ancillary activities were not getting the attention that they needed. Their turnover made them one of the biggest commercial enterprises in the country, and yet they were losing money. The potential of these activities was immense but was unlikely to be realised unless they had specialist management attention. The new Coal Products Division came into existence at the beginning of January 1963 and since that time has introduced new products, handled our important North Sea gas activities, and taken over the management of our new plants, including those producing smokeless fuel. Progress of this sort would have been impossible if the responsibility for ancillary activities had not been concentrated in this way.

Technology has presented management with some wonderful new tools, not least the computer. But these new developments have to be carefully watched. People can become over-enthusiastic about the potentials when they begin to grasp them. In my own industry, as in others, we had to resist the temptation to collect and store on the computer a lot of information that might, or might not, one day be used. Top management must be very stern about this and insist that only information which has a direct value is to be recorded in the data bank. And then there should be frequent reviews to test how often the information thus recorded is used.

Obviously, no business can be run without statistics. In the Coal Board we have done a great deal to ensure that we are collecting the right statistics, that they are then seen by the right people, and that we don't go on for ever recording particular statistics just because they were useful years ago.

I have, however, known people who have become completely obsessed with statistics. Some years ago one of the senior Directors in a coalfield whose results were the best in the whole country was being criticised because one particular statistic in his results did not show up too well. In exasperation he asked, 'Well, what do you want us to do? Get the statistics right or produce the coal?' It is the bottom line in the financial results that matters most. Did we have a surplus or a deficit from our operations? The senior official who was being roasted for his statistics was right. The bottom line for his collieries showed the end result everybody was looking for.

The re-organisation was a traumatic experience for the senior management, for none were unaffected by the change. Not only were many high-level posts abolished, but those holding them had to be fitted in, and in many cases this meant a move to another part of the country. It was a supreme example of team spirit, and two men no longer with the Board must be mentioned as having played a major part in the difficult task of carrying out the tremendous changes required.

One was Sir Humphrey Browne, then my Deputy Chairman, and the other Cyril Roberts, the Board Member for Staff. Both had served the industry since nationalisation with great zest and distinction.

Browne had been Deputy Chairman of the Coal Board for six and a half years and resigned in March 1967 to join the Board of John Thompson, the engineering firm, as Chairman elect. He was a mining engineer, highly honoured by his professional body, being one of the few men to receive the Gold Medal of his Institution, and an admirable backroom administrator. After we had taken our decisions on the form of the new management organisation we were to introduce in 1967, we set up a committee under his chairmanship to see the task through. This was work of great skill which was accomplished with his customary keen attention to every detail. In a way it was a pity he left the Board. The assignment didn't go all that well. When in fact John Thompson's was taken over by Clarke-Chapman in June 1970, Humphrey Browne severed his connection with the company, sought pastures new and became Chairman of the British Transport Docks Board.

He was considerably assisted in our re-organisation by Cyril Roberts, who took the lion's share in interviewing the personnel who were to be affected by the change. Within a period of eight weeks he interviewed hundreds of senior people. An old Etonian, he had inexhaustible patience and his courteous and friendly way of dealing with the personal problems of those he interviewed was an object lesson in individual communication.

Moving from policy to people was certainly not an easy task, but at last it was done and we all heaved a sigh of relief when the change-over date arrived and everything slid into place without too many mishaps.

Four years later one can see how important it was to make the changes, and whilst from year to year we quietly reviewed the position, nothing emerged that made any one of us want to return to the old method of management.

7
Communications—Within and Without

It was always necessary for me to be optimistic whenever I made a speech about the prospects of the coal industry. I was fighting hard to restore the morale and the confidence of the people on the payroll. But it was necessary to temper optimism with realism: I never hesitated to talk about the complexity and size of the problems facing us and I certainly always spoke out vigorously when I felt the industry was failing to do justice to itself or to realise its full potential. For example, at conferences of the National Union of Mineworkers I would often praise the industry for its productivity achievements, show how much more the machines were capable of producing if they were intelligently used, and then go on to show how absenteeism was holding us back from paying better wages, and how strikes were preventing us from keeping prices stable, thus preserving the size of the industry.

But it was always a difficult and delicate balance that I had to hold.

In using intensively the Press, radio and TV to help to get our story over, not only to the public (who are after all our shareholders), but also to our own employees (scattered as they are over the kingdom from Kent to Fife), contrary to the views of many people I was not seeking publicity for my own sake. I was quite sure, and have had it confirmed since, that the miners and other people on the Coal Board payroll felt good when they saw the Chairman standing up for them on the telly, heard him speaking out for them on the radio, and generally being acknowledged as a man of moment. It all helped to keep the industry big and important in the eyes of our own people and of our customers.

I did not find it at all difficult to adapt myself to the particular medium, or to the audience I was addressing. Experience in politics up to Cabinet level is pretty good training for exercises of this sort. But even industrialists who lack that background need have no fear of the mass communications media provided they put themselves in the place of the people watching them or listening to them. It is necessary to ask oneself whether the audience knows anything about the industry one represents. If the answer is no, then obviously one avoids technical terms. It would be no good, for example, if I used expressions like OMS (output per manshift), self-advancing supports (hydraulic chocks which, at the turn of a switch, take up their new position as the coalface advances), or 'inbye' (which means travelling in from the pit bottom towards the face) in a programme like *Any Questions*. But, of course, addressing a Colliery Consultative Committee I could use these terms freely.

Not that one should ever patronise the audience. Nor should one ever talk just for the sake of demonstrating knowledge of the subject or to create an effect. It always shows. To talk naturally and easily is the best way, though I know that this is difficult for some people. Affectation, pomposity and arrogance will soon be exposed in a television interview, but it's difficult to be pompous, affected or arrogant if one is relaxed. So that should be the aim all the time.

I have always been eager to use the most up-to-date technical methods of presenting a speech and I had plenty of good chaps to help me do it. One of the most effective of these attempts was at the annual conference of the British Association at Cambridge in September 1965. I was speaking on the subject of the potential contribution of coal and in order to impress the scientists present with the fact that we were out to utilise all the scientific aids that we could, I cut into my address to set up a live television link with the manager of Bold Colliery at St Helens in Lancashire. The manager was able to talk direct to the delegates via the cinema screen behind me, and then I described the various items of equipment that were shown. Finally, amid tension that could be felt, I said that I was now going to press a switch on the rostrum beside me and the audience would see the face conveyor three-quarters of a mile underground, at a pit well over one hundred miles away from me, stop and start. To the relief of the technicians who had set up this demonstration, it worked. And there is no doubt this spectacular presentation brought home the points I was making about the technical advance in the industry.

'If you were spending your own money, would you spend it in this way?' This is a question I have asked many times in the last ten years. Of course, we have more scientific ways of measuring the expected results from a project, like the discounted cash flow method, but still, my question has often been a useful one.

One of the first people I ever put it to was Geoffrey Kirk, who had recently been appointed Head of the Board's Public Relations team. We were talking in my office during the time that I was Deputy Chairman and we were discussing *Coal Magazine*, which the Board were then publishing. It was quite a handsome production—a photogravure job—but I guessed that it wasn't doing a real job. There was very little local news in it and the circulation, at a time when there were still 600,000 people in the industry, was only about 30,000 copies a month. My first contact with it was over a decade earlier.

When I was first appointed Parliamentary Secretary at the Ministry of Fuel and Power, I had been interviewed by a reporter from *Coal Magazine*. Some months later (well after the evaporation of any news value my appointment may have made) I was sent a complimentary copy of the issue concerned. I remember thinking then, as I read through a series of flat and totally unprovocative feature articles, that it had about as much relevance to current mining problems as the *Tatler* magazine. When I read the current issue shortly after coming to the Board, I saw that it had not changed. Any miner reading it might have been forgiven for thinking that everything in the garden was lovely, and that an insatiable market was waiting to gobble up whatever he produced, irrespective of its price or quality. It was a good thing so few miners read it.

I had not represented a predominantly mining constituency in Parliament for over fifteen years without knowing that mineworkers were intelligent people who deserved to be well informed. It seemed to me that the employer had not only a moral responsibility to see that they were, but a commercial interest as well. It is quite pointless in any big organisation to talk grandly about equipping management with an understanding of business objectives unless the rank and file of workers and first-line supervisors know not only the job they have to do, but the relationship which their work has to the enterprise as a whole. In an organisation which has to adapt itself constantly and speedily to marketing forces, this consideration is vital.

Kirk agreed that he wouldn't spend his own money on *Coal*

Magazine and we decided to switch instead to a tabloid-size newspaper which would be published in local editions. That paper, *Coal News*, has for years now been recognised as the best in its field. We recruited an industrial correspondent called Norman Woodhouse from a provincial daily newspaper and we told him that he was to run it strictly as a newspaper. Anything that got into *Coal News* must be there on its news merit. We would give full coverage to Union views and activities. There must be a home page and features for children, because we wanted the men to take the paper home. The formula quickly proved to be extremely successful and for some time now sales have been running at the rate of about 65 per cent of the total labour force. This is excellent, especially when one remembers that mining is an industry where father and son and brothers often work at the same pit, so that one copy will be read by two or three employees.

Initially, it is fair to say, opposition to the paper came not from the Union but from top management. The ruffled feathers of bureaucracy produced complaints about inaccuracy and 'disturbance at *Coal News*' failure yet again, to consult Departments'. Occasionally the line the paper took on certain issues irritated me, but I managed to resist the temptation to say so to the Editor, who was a professional and knew how far he could go. When I told him (as I did when we offered him the job) that *Coal News* was not to be a 'boss's newspaper' he knew nevertheless that he was not expected to produce a headline 'Robens is a nit', any more than the Editor of the *Daily Express* would be expected to run a series of articles extolling the virtues of the Common Market. Equally he knew that his paper would rapidly become discredited (and thus lose circulation) amongst its readers unless it gave fair coverage to union activities and union opinions.

Coal News has proved eminently successful in steering a course in such perilous seas. It prints the facts. But basically it prints news about mining and mineworkers. To miners, the promise of Drax B power station for coal is news no less than the athletic achievement of Dorothy Hyman. A picture (house magazine-style) of my wife and I 'sharing a joke' with the Mayor and Mayoress of Middlecombe-on-Sludge wasn't news, so it didn't get in the newspaper.

During wage negotiations *Coal News* ensures that the Board's offer is factually made known to the rank and file of the industry. In doing so, it frequently incurs the wrath of the Union because its communications with its readers are often swifter than those of the Union to its members. The paper has had more lives than the proverbial cat. Various areas of the NUM have threatened to ban its distribution. Some top managers in the Board are frightened by its 'journalistic irresponsibility'. But the paper lives because 210,000 miners pay 2p a copy for it a month. No house journal could have any higher justification.

The significance of *Coal News* is that it gave us a really effective medium of communication with our workpeople. It used techniques that had been successful in the newspaper business. In effect, we were presenting news about the industry to men in a way in which they are accustomed to get their everyday news.

Why should the Union leaders—or some of them, at any rate—so dislike *Coal News*, especially when their members can read far more about the Union's activities and views in our paper than they can anywhere else? The answer simply is that they realised we now had a better means of communicating with their members than they had themselves. We were often in trouble because we published in *Coal News* the details of wages offers. Before the paper existed, the Branch delegates would go back to their local meetings almost carrying in their hands the increased wages that they had won from the Board. Now the men got a permanent record, in detail, of what had been agreed. In the days when the flow of information was dependent on the delegates, the facts and figures would be given orally at a Branch meeting, perhaps with only a very small proportion of the members present, and there could be misunderstandings and arguments about the information issued there. With complicated new agreements, we often published in *Coal News* a pull-out guide. At one pit where the Secretary of the Union Branch was clamouring for a ban on *Coal News*, he was nevertheless walking round with a copy of such a guide tucked in his pocket and, whenever he got a query from one of his members, he had no hesitation in pulling it out and referring to the despised paper.

Whenever we were criticised, my reply was always the same. 'They may be the Union's members, but they are our employees. We have every right to communicate with them. And we have every intention of going on communicating with them.' No employer at any time should ever abrogate his right to communicate with his own workpeople. At the same time the trade union is entitled to be provided with reasonable facilities to communicate with its members. This we certainly did. The Union did not even have to collect the union dues; we deducted them from wages, with the authorisation of the employee of course, and handed over the sum collected in bulk. Arrears of contributions are not something the Mineworkers' Union has to worry about.

When the delicate matter of wages is being discussed it is imperative that every employee should be fully aware of all the facts. The mineworker is entitled to have, not only the details of an offer on which he may be required to vote, but also the reasons why the Board are unable to offer bigger increases, what the effect on prices is likely to be, how much business we may lose to the competitors, and how many miners' jobs may be lost if increases are not covered by rises in productivity. This information is needed before the man can make a sensible decision about where to put the cross on his ballot paper. There was no evidence at all that the rank-and-file members of the Union resented our attempts to keep them informed. On the contrary, the popularity of *Coal News* and the reception that has been given to our other attempts to improve the flow of information show that most people welcomed what we were doing.

Some may claim that this is going over the heads of the Union leaders, but I deny it. The Union must improve its own methods of communication, and I welcome the great improvements the National Union of Mineworkers has made.

While we and the more alert Union leaders used *Coal News* for putting across our views and policies, there was a very lively letters page which enabled the people in the industry to say what they thought of us. And some of them did this in very vigorous terms. Whenever I had a letter that was critical or even abusive, as one or two of them were, they were published, with my comments underneath them—always, of course, most courteous and free from any trace of resentment.

Coal News was the forerunner of a whole series of periodicals, each catering for a specialist audience. The senior staff had their publication *Management News*. For under-officials (they would be called supervisors in other industries) we produced *Inbye*, while *Coal Products Newsletter* looked after the particular interests of our employees in that Division.

These periodical channels of communication (*Coal News* is monthly, *Management News* and *Inbye* appear every other month, and *Coal Products Newsletter* comes out two or three times a year), are supplemented by special issues on particularly important occasions or topics. Writing about industrial relations earlier, I have shown how we used direct mail methods to ensure that every man in the industry knew exactly what was going on in the negotiations. And we have also used local press advertising in unofficial strikes when we wanted everybody to be clear about the issues involved and their effect on the jobs of other people in the industry.

I have always enjoyed good relations with the newspapers, and particularly with the specialist writers like the Industrial Correspondents, making myself available on request for interviews and press conferences in the coalfields. This was also a good way of communicating with the people of this still vast industry and with the colliery village communities. It was not only the men who worked in the pit who were interested in its future; on the payroll of 2,000 men at a colliery, about 5,000 people depended.

Some of the Industrial Correspondents have been among the most distinguished people in journalism since the War. I could quote many names but two in particular come into my mind. Sir Trevor Evans, now retired, but for many years the Industrial Correspondent of the *Daily Express*, was particularly knowledgeable about the mining industry—as indeed he might be expected to be. Trevor's first job when he left school was at a South Wales colliery and his father was killed in a colliery explosion. Nevertheless, he always had warm feelings towards the industry where he might have been excused bitterness. But bitterness was foreign to Trevor's personality.

Geoffrey Goodman, now Industrial Editor of the *Daily Mirror*, but formerly with the *Herald* and *News Chronicle*, is another man whose reporting on coal has always been penetrating and fair. Both these men knew that what they were writing about was not just an industry but, above all else, people whose standard of living depended on its prosperity. Because of their backgrounds, both could readily identify themselves with working people, although their understanding of what management were trying to do was also complete.

Largely I believe because of the influence of Trevor Evans, the *Daily Express* has always dealt fairly with the industry. It may have criticised the Coal Board but rarely has it turned its guns upon the miners. In fact, contrary to most people's impressions, no newspaper has been more ready to praise their efforts and rejoice with them in their successes.

Curiously, its stable mate the *Sunday Express* has pursued a completely different policy and has repeatedly made the most bitter personal attacks on me.

There was one period when almost every week the *Sunday Express* would attack me, suggesting that I was not competent to run the enterprise. I must say that I found this public attitude difficult to reconcile with the invitation to me, which I accepted, to make the presentation at a private function to Bill Needham, one of the Directors of the Beaverbrook Group who was retiring. After the speeches and the hand shaking, no one was more personally effusive in his gratitude to me than John Gordon of the *Sunday Express*. Why his organisation should have bothered to get a man whom they suggested week after week in the columns of the paper was incompetent to do his job, to come and do the honours to one of their retiring colleagues, has always puzzled me.

But no matter. I was always on good personal terms with Max Aitken the Chairman, and readily accepted an invitation from Brian Nicholson, one of his Directors, to become President of the Communications, Advertising and Marketing Education Foundation, in which he was specially interested. So why should I worry about the journalists who appeared to dip their pens in bile? It at least proved one thing to the credit of the *Sunday Express* and that was that they had complete editorial freedom and could take the pants off a friend of some of their Directors without managerial interference. My legal friends told me that I had more than once passed up a good case for libel, but I always replied that 'today's newspaper is tomorrow's fish and chip wrapper'.

The only other man I knew who never permitted newspapers to ruffle him was Clem Attlee. That was because he only read *The Times*, did the crossword and was quite impervious to the popular press comment. If anyone drew his attention to a particularly vicious piece, he shrugged his shoulders and snapped, 'Not interested.'

One of the traditions started in the first year of nationalisation was for the Board to give a luncheon to the Industrial Correspondents and Editors of technical papers as early in January as the statistics for the previous calendar year were available. I used to review the results, discuss prospects for the future, and answer questions before lunch. While the agencies and evening paper reporters were telephoning their stories I would do any TV or radio recordings that were needed and then, business over, we would all sit down to a steak and kidney pudding lunch. This menu was the idea of Henry Donaldson, the Coal Board's Chief Press Officer, who argued that the Press got sick of turkey and the other Christmas fare when they were entertained at this time of year. These were always happy and often hilarious occasions and they, above all, yielded enormous press coverage.

I was only let down once in the ten annual lunches I had. I was discussing the activities of a senior civil servant and one of the journalists (a very well-known one at that) broke the rules by relaying a comment of mine, somewhat out of context, to the man concerned. The next day the civil servant rang me up. He was furious and gave me no time at all to explain that he had been 'conned' with a favourite trick of some journalists in an endeavour to get further information. He rattled on at a great rate and then cancelled a dinner appointment he had with me and said that all further contacts between us must be on a purely formal basis in his office or mine. His ban didn't last very long, as he retired shortly afterwards. The joke was that within three or four weeks of the incident, the word went out from 10 Downing Street that this particular journalist should not be given any personal interviews by Ministers or senior civil servants. He had tried something similar again but this time quiet judgment was used instead of emotional rage. Incidentally the senior civil servant never ever told me what the journalist claimed I said.

But for all that the annual review and steak and kidney lunch was an event I always looked forward to. I genuinely liked the journalists and I had the same regard for the Industrial Reporters that I had for the Lobby Correspondents when I was in the House of Commons.

They were all hard-working, nearly always seeking and writing their copy against the clock and, provided you were on the level with them, they rarely let you down. Certainly in all the years of my public life, I have been bitterly criticised and fairly praised, but I have rarely been let down and on more than one occasion I have been greatly helped. Equally, I never let them down either. The greatest compliment paid to me in this direction came from the Chairman of the Gallery Correspondents, the late Edgar Hartley of Thomson Newspapers, who at a lunch they gave me when I left the House said: 'There is one thing about Alf Robens that has made him a friend of all of us, and that is, if he felt he could not disclose something he said so frankly, but he never tried to mislead.' I was determined to ensure that the standard was maintained.

At my first solo appearance at the Coal Board's New Year lunch I gave broad hints that, because of increased productivity in the preceding twelve months, the Board, who were due to meet the NUM on the following day for wage negotiations, would be able to make a reasonable offer. I was also able to forecast that the deficit, after meeting interest charges, would be down by about £6 m. by comparison with the previous year. I said that we could break even in 1962 with a good deal of luck and freedom from industrial dispute, which in the event we did. Coming after a series of heavy losses, this was news indeed.

Broadcasting, both on television and radio, I have always enjoyed. Many people have asked me why, in a very crowded schedule, I was willing to make time to appear, for example, on programmes like *World at One*. The answer is that I usually managed somewhere during the broadcast to get in a plug for the coal industry. Making speeches or being interviewed on TV and radio were always easy for me. Addressing an audience is, for me, what a round of golf is to many men—a form of relaxation.

The provincial Press were always a good crowd to be with. Their reporting too was very objective and their personal likes and dislikes did not come peeping through in their pages. A few weeks before I retired I was given a tremendous surprise which took me aback quite a bit. I was asked when planning a visit to Yorkshire to allow time for a farewell drink with the journalists who had covered my frequent and sometimes controversial visits to that turbulent coalfield. It is rare for pressmen to become attached to businessmen, yet the Yorkshire journalists gave me a silver tankard and tray. It was inscribed with their thanks, and I shall not forget the very pleasant things they had to say, one of which was that I had given them some good copy over the years.

Whenever I went to a pit we would always invite the local Press so that the miners in the Area, even if they hadn't seen me themselves, would read reports of my visit in the local papers. These pithead conferences were always well attended and very often a local journalist would scoop the national pool as a result of a well-timed question.

It took some of our local management a little while to cotton on to this. I had been down one pit with the Area boss, and when we came back to the surface the radio and Press were waiting. The Area Director assumed that I would have my bath and change before I saw them but, of course, he hadn't realised that the Press would be much more interested in getting pictures of me with a miner's helmet on and a mucky face. So I got straight on with the interviews, and when he came back spruce and rosy faced from his hot shower bath, he asked, 'Well, shall we have the Press in now, Chairman?' But the press conference was over and he had missed the boat.

Moving about the coalfields was stimulating and rewarding and there is no doubt that the desk-bound chairman of any industry is at a great disadvantage in not being able to make the names of people in his organisation come alive. Nor is he able to visualise situations that arise in activities or parts of activities that he has not seen. Getting around, seeing for myself, discussing and maintaining a continuous dialogue with men and management provided a sparking point for all sorts of ideas.

At the same time the daily decisions based on memoranda by the Board's advisers had to be dealt with promptly. A twenty-four hour delay at Chairman level held up action by a large number of people for a much longer period. But I was lucky in that travelling was never distracting, so it did not matter whether I was in a plane or train or a car, the old briefcase was my companion, which is why today I never know the way to anywhere. I very rarely looked up.

So I have been anything but a desk Chairman. Almost every week I spent two days out of London, visiting pits, discussing with Area Directors their results and their problems, going to see important customers, visiting our research establishments and so on. Even during the days I was in London, I had very little time to clear the paper that came into my office because I would be involved in meetings, either in Hobart House or in other parts of London. For the whole ten years I met my Directors-General (Heads of Departments) every month for what I called our 'informal discussion'. Every day there would be a business lunch with either Board people or people connected with the industry. So there was little time for the inevitable office work—certainly not during office hours. Most weekday evenings I had a dinner appointment, often involving a speech. But despite this I made sure that there was no hold-up in administrative decision-making at my level. No matter how late I got home, I would settle down with my bag of work and clear everything before I went to bed that night or—more often—in the early hours of the next morning. In this way, people were not kept waiting for decisions. Almost always if they got a paper to my office before 6 o'clock in the evening, they would get a decision the next morning.

Whenever I returned from a visit out of town in the Board's aircraft, a car would meet me at the airport with a bag and I would spend the time while being driven to the office in working through all the correspondence that had come in and indicating what was to be done on each letter.

My secretary, Miss Doris Proctor, whom I inherited from my predecessor Jim Bowman, a quietly efficient, non-possessive, un-fussy personality, could almost tell by my handwritten scrawl what kind of a vehicle I was in and the speed it was going. I dismissed the idea of a dictating machine. You could always tell a letter that had been typed from one—usually far too verbose and always unnecessarily long. I like both to receive and to send short letters, which deal with the main points and leave out the trimmings.

I also had a very good private office, through which passed a succession of excellent staff officers. Most of them had joined the Coal Board through the industry's Administrative Assistant Scheme. We recruited arts graduates and gave them about three years' thorough training in different departments and at different management tiers within the organisation. Most of those who worked for me spent two or three years in my office. It was strenuous work for them, and when they left to go on to other departments I'm sure they had benefited from the experience of having worked at the centre of the organisation. They have all now won their way to senior positions in the industry, and the organisation has gained from the contribution of these able young men who have worked very closely with me.

A typical programme for my coalfield visits would be to leave the office around 5 o'clock in the evening, drive to Luton, where we kept the Board's aircraft, and fly on to the nearest airfield to my destination. In the evening there would probably be a working dinner with local industrialists. Next morning I would be at a pit. After a quick run over the colliery's plans with the local management, I would change and be off underground. After travelling perhaps a couple of miles on a man-riding train or belt conveyor equipped for the purpose, there would follow a walk of half a mile or so along the roadways perhaps 14 feet high.

Once at the face it was a question of crawling on hands and knees for a couple of hundred yards because the average roof is only 3 feet 6 inches from the floor. From the other end it would be another fifteen-minute walk back to the manrider, then back to the pit bottom and up to the surface in the cage.

This would take getting on for a couple of hours and I would spend half an hour with the Press. After an underground trip a hot shower bath is a grand experience. It left me ready for a buffet lunch and drink with the pit's consultative committee.

During the afternoon there might be an accountability meeting with the local Area management, reviewing their results and discussing their problems and plans.

Then about 5 o'clock it would be a car to the nearest airfield, where the gentle Dove (about which more later) would be waiting to fly me and my companions back to London.

Throughout the day my Staff Officer would be passing on to me telephoned messages from Hobart House so that queries could be settled swiftly.

Sometimes when I have been staying overnight in the coalfields for a business meeting the next morning, I have left my hotel and paid a surprise visit to any nearby Coal Board premises, not with the intention of catching people out, but simply to make the most of an opportunity to see all aspects of the Board's business. After the initial surprise the people on the spot were always tickled to entertain the Chairman and to tell him all about their jobs. In this way I have visited Coal Board brickworks and by-products plants, as well as collieries. Once or twice I have found the security arrangements to be less effective than they ought to have been. One Sunday I was driving past our show-place colliery in Nottinghamshire, Bevercotes, and turned in at the gates and straight up to the manager's office. There was no one about, so I took one of the pit plans and put it on his desk with my card saying I had enjoyed the interesting visit. The end result was that they greatly improved the security.

And of course, the grape vine was soon in action; the story grew and there was a general checking of security all over the Area.

People sometimes ask me if I have a memory system, or whether I have a photographic memory. I find I can remember people pretty well and also recall events in some detail, even though they took place a long time ago.

But I certainly have no system for memorising things. I think it must be that I find it possible to concentrate completely upon what I am being told, or on what is happening. I can shut out of my mind everything except the immediate subject. As a boy in the church choir I used to sing the old and lovely hymn 'Lead Kindly Light', which contains the words: 'I do not ask to see the distant scene; one step enough for me.' And that's what I've tended to do and it's not a bad idea in business life—to concentrate all one's knowledge, experience and skill on one thing, and one thing only at a time. When you turn to the next task you can wipe the previous one completely from your mind. The ability to concentrate also means you are more relaxed. It must take a lot out of a man to be worrying about three other things in addition to the one that he is trying to settle at that moment. In my view, single-mindedness equals relaxation.

When the Coal Board acquired the de Havilland Dove aircraft I have mentioned, a number of eyebrows were raised. One newspaper in its very early edition condemned me for it but, having seen the other newspapers' welcoming comments, in its later editions said it was a good idea. Executive aircraft were not very numerous in Britain in those days, and that one of them should be operated by a nationalised industry obviously created a great deal of surprise.

One miners' MP, Harold Neal (a former Parliamentary Secretary) who sat for Bolsover, was scathing about it. However, the decision had support from some unexpected quarters. Jim Hammond, the Lancashire miners' leader and a lifelong Communist, said: 'If it were necessary I would not mind if Lord Robens travelled about in a Sputnik . . . If he can do the job better with an aeroplane by all means let him have one.'

Of course, it was always described in the newspapers as Lord Robens's own aircraft, but in fact it was available to be used by officials who were travelling about the country on Board business, and we worked out a utilisation system that enabled even the most junior members of the management team to travel on it that quickly made our little six-seater (registration G-ARUM) the most intensively flown private aircraft in the country. I always enjoyed seeing a new face aboard and took every advantage of a talk and drawing the chap out.

The Dove has certainly proved its value as a management aid. I and many hundreds of other Coal Board people have been able to get through more work and spend more nights at home than we would have done without the aircraft. Without it we couldn't have had a single Regional Chairman (Wilfrid Miron) for the Midlands and South Wales. It has also been proved to be financially very economical.

We had many amusing experiences with the Dove. Sometimes, we were able to use RAF fields when they were more convenient for our destination than civil airports. One afternoon I and a group of others were flying to a management conference in Scarborough. It was a Sunday, and when we landed on an airfield near the town, I was touched to see as we taxied to a halt that a very senior RAF officer and his Adjutant had had the courtesy to turn out to receive us. Wanting to show my gratitude to him, I plunged towards the aircraft door saying: 'There's the Station Master, I'll go and thank him.' An alert colleague in the aircraft just managed to grab my coat tails and hiss: 'Station *Commander*, Chairman.'

Not everyone enjoyed travelling in our litle Dove, especially if it was the first experience in a small aircraft, and Wilfrid Miron tells the story of when he and Mrs Miron were travelling back with me from Hucknall to the aircraft's base in Luton, in order to attend an official function in London that night. I myself was due to make a speech at a City dinner that evening. It was obvious to me that Mrs Miron was, to say the least, pretty uncomfortable, so away went my papers. Having spotted her unease I managed, throughout the trip from Hucknall to Luton, to talk to her about feminine matters—dress, fashion and such like—in order to take her mind off the aerial environment. The flight from Hucknall to Luton is not a very long one and I think she would have been quite ready to fly on for much longer in order to continue our conversation. My own wife is not a good traveller in small aircraft, so I was able to sympathise with Mrs Miron's discomfort and use the technique I had found effective in my own wife's case.

The Dove was one of the first British aircraft to use the Berlin corridor after the Russians had sealed it off. I was going to Berlin to inspect a new type of kiln which we were considering using in some of our brickworks. (The Coal Board have for many years been among the three biggest manufacturers of bricks in the country, mainly because of the interests sensibly built up by the former colliery owners. Many thousands of bricks were used underground in collieries and the fireclay needed to make them was often worked in getting coal.) We had applied for permission through the Board of Trade and our pilot, Basil Allom, a former Fleet Air Arm flyer who has now been followed into the commercial flying business by his two sons, attended a special navigational and political briefing.

We were coming out on our way to Düsseldorf when reports apparently reached London that our aircraft had been shot down by the Russians.

In fact the aircraft shot at and forced down by them was being piloted by Hughie Green, the TV personality, who was on his way to do a *Double Your Money* programme for RAF people stationed in Berlin. Evidently there must have been some slip in notifying details of his flight to the Russian authorities. Hughie quips that perhaps the Russians had seen the programme, but he was under fire for about forty minutes and his plane was held for three weeks.

When we landed at Düsseldorf I found myself in the position of answering the phone and, like Mark Twain, saying that the report of my death was an exaggeration.

A good deal of travelling by senior staff between London and the provinces was reduced by the gradual movement of staff from London into the coalfields. At one time the Board had seven roofs in London: today they have only one—Hobart House, and even part of that is rented by ancillary undertakings.

For some time now Governments have pleaded for businesses to move their Headquarters out of London. In January 1966 we announced that we should follow this policy as far as we possibly could. We transferred to Doncaster all Headquarters staff who were not essential in the capital. The whole of Purchasing and Stores Department, most of Production Department, Accounts, Pensions, and Organisation and Methods went to Doncaster. Coal Products and Research and Development moved out of Central London to Harrow. This enabled us to close all Coal Board offices in London except Hobart House. This was another big move involving very large numbers of staff that was carried out smoothly because of the care and trouble that were taken over it by those responsible.

The Coal Board was the only publicly owned enterprise which was truly national in character, unlike the gas and electricity industries (which had largely autonomous area boards), as it controlled absolutely and completely the whole of the industry in the United Kingdom and this had its political complications. Even those splendid Welshmen and Scots who prided themselves on not being nationalists only wanted a slight scratch to show that nationalism lay just below the skin. Any matters therefore which affected Scotland and Wales had to be very carefully considered from a political point of view as well as an economic one.

One example of this was the impact of the Board's pricing policy. For some years there had been a fair amount of equalisation of pit prices between coalfields, with the result that they had become so out of line from the true costs that pits could easily have been closed for the wrong economic reasons.

During the spring of 1962 we introduced price increases on a selective coalfield basis. The reasoning behind this was that the profitable pits in the Midlands were having to charge higher prices than would otherwise be necessary in order to carry the losses being made in other coalfields. So we increased prices in Scotland and Lancashire. These coalfields had been running at a heavy loss for many years. There wasn't a 'miff' from Lancashire, but plenty from Scotland.

It was one of the most controversial decisions to come out of the industry and was described as penal and a selective imposition on Scotland. This was despite the fact that the housewife in Edinburgh and Glasgow got her coal cheaper than the housewife in London, as indeed she does now. Protest meetings and scores of hostile letters—you would have thought that the 'hammer of the Scots' had come back. To this day there is still resentment north of the border about the 'selective' price increase.

I was undoubtedly a very unpopular man in Scotland because all these matters were inevitably personalised. The Scottish Nationalists in particular resented what they described as London remote control. However, when I suggested that the Scottish coal industry should be set up under a Board independent of the NCB, the criticism sharply diminished. Accumulated losses in Scotland then stood at £120 m., whereas the deficit on the industry as a whole since nationalisation fourteen years earlier had been £90 m. In other words, the rest of the industry would have been comfortably in the black had it not been for carrying the losses of Scotland.

The selective price increases apparently aroused the ancient fears of the National Union of Mineworkers leaders that in some way they represented a step-back towards the pre-nationalisation practice of local wage agreements. So in my speech to the Union's annual conference that summer in Skegness I went out of my way to explain the Board's policy. I said that we were looking to the Yorkshire and Midlands coalfields to earn enough profits to carry the burden of the coal industry's interest charges which, at that time, were running at more than £42 m. a year. But, with increasing competition from oil, a lot of the business those coalfields were doing was in danger. This was why we could not allow losses on the other coalfields, which in some ways were more difficult to work, to jeopardise the continuing prosperity of these coalfields which were, and would always be, the vital heart of the industry. If the coal produced in these Areas could not be sold, then there would be no hope for the rest of the industry.

Because of the very serious financial position of the coal industry in Scotland about this time I was giving a large slice of my time to its problems. The Rothes colliery in Fife, a new mine sunk at great expense, had proved to be a tremendous loss-maker. Commissioned in 1946 just before nationalisation, it rated for compensation to the former owners but it turned out to be a poor asset. It had been visited by the Queen, and the new town of Glenrothes depended to a great extent upon the wages of the 2,670 men whom the pit was expected to employ. At its peak it employed 1,340 men. The project cost £9 m. and had operating losses of £3·7 m. Struggle as men and management did they couldn't make a go of it.

The reserves the pit was supposed to work were very heavily waterlogged and badly faulted and even in some places completely absent. I don't think anyone believed that the Coal Board would have the courage to shut this pit down, because it had always been regarded as the show-piece of the Scottish industry. When a report showed that of the sixteen coalfaces opened on one level since 1957, fourteen had had to be abandoned though they should have produced about 2,500 tons of coal a day, I realised that this unpleasant decision must be faced.

The pit was closed, but not before a tremendous battle in which everybody joined in and grim forecasts about the future of the town of Glenrothes were made. It is now without the basic industry it should have had and its development has been slower than it would otherwise have been. Unemployment there is extremely high.

At the time, what with the selective price increases, the closure of Rothes and the general air of gloom about the future of coal in Scotland, it was extraordinarily difficult to maintain the morale within that coalfield. But we persevered and we subsequently got a break when we secured the huge Longannet Power Station for coal. These breaks didn't come very often and when they came we made the best of them.

Because we had built up an internal communications system second to none, we were able to explain our difficult decisions like the selective price increases and the Rothes closure to all our own people. And we were also able to publicise our triumphs like Longannet.

The Dove and the coalfield visits were other ways of pulverising the distance between the man at the coalface and Hobart House.

Because of the work we did on communications we were able to weld together the people of this huge industry to an extent unequalled before in mining or any other industrial organisation of any size. We continued to have our internal arguments of course, but to outsiders—Ministers and customers—we were powerfully united in a common purpose. Which is why the industry has survived its vicissitudes in good health and good heart.

8

Mixing it with Ministers

One of the great weaknesses of Government, that has been borne in upon me more and more as the years have gone by, is the frequent change of Ministers and senior civil servants which has made the task of those responsible for running state enterprises extremely difficult. No private or public business could possibly escape bankruptcy if the men at the top were changed so frequently.

In the old days up to the end of the 1930s, it didn't matter very much. Intervention in the economy's development by Ministers (and by the Government as a whole) up to that point was small indeed. Perhaps the mass unemployment that existed for twenty years was occasioned because Governments did not interfere enough. Be that as it may, the position of the Government and the economy is very different today. Over 50 per cent of the nation's total capital investment is in their hands. What is more, it is the vital part of the investment, which in turn triggers off investment in other industries. Nothing has been more ridiculous than the spectacle of Ministers complaining and worrying about the lack of investment in the economy as a whole, while at the same time the Treasury is engaged in holding back investment because money borrowed is raised by taxation and all Governments want to lower taxes rather than raise them. When it is suggested that money for many nationalised projects could be raised directly from the City, this is rejected, very largely I suspect to retain control. The grip of Government upon national-ised industries gets tighter each year. It is obviously thought that civil servants are more capable of running the state enterprises by some unique method of remote control than are people appointed by their Minister under an Act of Parliament. They may well be right. But if so and if all the wisdom and knowledge lie in the senior civil servants, then the Minister should appoint civil servants to run the particular industry or bring the public enterprises into the Government Departments, as used to be the case with the Post Office. Neither of these is practicable, so we are left with the present most unsatisfactory position of very short-stay Ministers and somewhat longer-stay civil servants, both with authority to make decisions.

Governing in the 1970s with the tarted-up machinery of the 1870s is certainly not the way to administer efficiently a highly industrialised country like the United Kingdom. That is why a few years back I advocated substantial changes in the procedures of Government so as to ensure the maximum use of resources as one would in a business. The idea became known as Great Britain Ltd. It would provide, in my view, a more efficient use of capital, the elimination of needless bureaucracy, a general smartening-up of the Civil Service itself (which many civil servants would like), and a dividend declaration that would bring a bright smile to the face of every citizen. This concept got nowhere of course. How could it? It trod upon too many of the corns of politicians and civil servants.

I was once asked by Barbara Castle if I could suggest any name for consideration for the Chairmanship of the Board of British Railways. I nominated Sir Matthew Stevenson, who had been a member of the Stedeford Committee along with Dr Beeching (as he was then), which led to the railway re-organisation and the subsequent appointment of Beeching to the railway chair. Sir Matthew was at that time the Permanent Secretary of the Ministry of Power, and he certainly thought he knew how to run the coal industry better than I did, as his advice to Ministers showed. It seemed to me that it would have been an interesting experiment to allow someone who had tried to practise remote control the chance to show what he could do when made directly responsible. He wasn't appointed, but Bill Johnson, a man who had been with railways all his life, got the job and this seemed a sensible and practical decision.

During my ten years as Chairman of the Coal Board I had to deal with no fewer than ten Ministers. Yet on the management side there has been much more continuity: in the twenty-four years since nationalisation the Board has had only five Chairmen.

One would have thought, with the lack of continuity on the Ministerial side, the views of the Chairman would be treated seriously. This was not my experience except during the period of the Conservative administration up to 1964. With the advent of the Labour Government the attitude changed. This was

something I found difficult to get used to. I knew politics. I had
served in the Labour Party and in the Government of Clem Attlee
in the immediate post-war years alongside all the members of
the Wilson Cabinet and half the rest of the Ministers. There was
in fact no one in the Labour Government of 1964 from the Prime
Minister down who had had anything like my experience of
nationalised industries. So it seemed to me these were added
reasons why my views should have been carefully considered by
former colleagues.

I gave them all credit for wanting to see nationalisation succeed;
I believed that they all were passionately concerned about the
impact of Government decisions upon the well-being of people;
and I thought they understood that the miners had created the
Labour Party for them and were the most steadfast and loyal
group of supporters that existed.

But it was not so. The very same people who in 1971 in Opposition were screaming their heads off about the loss of jobs in
Rolls-Royce and Upper Clyde Shipbuilders, had forced through
policies that reduced the jobs in mining by over 200,000 and
cost the country millions of pounds in abortive capital investment. So I had to fight their policies every inch of the way,
though it wasn't pleasant for me to have to do so. The mining
communities showed their displeasure by staying away in their
thousands at the 1970 election.

I suppose one could have more easily coped with the frequent
change of Minister if they had all had the same competence and
interest, but this was far from being the case.

I was really sorry in October 1963 when Richard Wood, after
four years as Minister of Power, was replaced by Freddie (later
Lord) Erroll. Wood had been a good, compassionate and understanding Minister; but his successor, who had been President of
the Board of Trade, regarded himself as having been demoted
and, during his short spell at Power, made no impact whatever on
me.

Richard Wood was a man with whom you could work well
because he was not only kind but a very patient and exceptionally
courteous man. Differences in views or emphasis were expressed
in a very civilised manner, there was never the arrogance which
power brings to most people. I knew him, too, in another
capacity and that was in his work for disabled people. I joined

him in 1951 on the Board of Governors of the Queen Elizabeth Training College for the Disabled, a residential establishment which trained disabled people for trades and professions so that they could once again earn their living in the world. Wood is badly disabled himself, although a casual acquaintance would never realise it. He lost both his legs as a very young man in 1943 when serving in the army. This disability never stopped him from visiting pits and travelling underground, though it must have been very arduous and difficult for him. The two years with Dick Wood were the easiest of the whole decade. It was during his reign that the coal industry got the most effective protection that ever came its way.

I had a number of private conversations with the present Speaker of the House of Commons, Selwyn Lloyd, who was then the Chancellor of the Exchequer. The outcome was the announcement in his 1961 budget speech that he was going to tax fuel oil at the rate of an extra 2d. a gallon. There was an immediate and bitter storm of protest. Everybody assumed that this tax had been levied to help the coal industry. Certainly this addition to the price of oil improved the comparison between coal and oil by nearly £1 a ton in our favour. The storm was so widespread and intense that Richard Wood, three days after the budget speech, found it necessary strenuously to deny that the tax had been introduced for the sake of protecting coal. In fact this accusation persists even now, ten years after the event.

Selwyn Lloyd stoutly maintained that the imposition of the 2d. a gallon was for revenue-raising purposes, and I was quite content with that. I knew that a number of senior Treasury officials were strenuously against it. Because he was a Tory Chancellor, the miners never gave Lloyd the credit he deserved for the best lift to the coal industry we had ever had since the strong competition from oil began.

In the run up to almost every budget since the tax was imposed there have been strenuous attempts to persuade successive Chancellors to remove it, but so far without success.

Freddie Erroll appeared to have had all the stuffing knocked out of him when he was transferred to Fuel and Power from the Board of Trade. He was an experienced administrator, but I cannot recollect any fresh initiatives coming from him. I remember that John Raven, who was then the Director of the

British Coal Exporters' Federation, invited Erroll and me to dinner to discuss, amongst other things, export possibilities. It was a pleasant evening: after all, Freddie Erroll was an affable and genial man, with a fund of interesting experience—a good table companion. But he had nothing to offer by way of advice to John Raven and me on export policy. I remember his words very clearly: 'As far as coal exports are concerned, you must do that which you regard as profitable and commercially sound.' It was of course very good advice, straight out of the text book.

Erroll once told me that it was he who suggested to the Prime Minister the line 'exporting is fun' which Mr Macmillan used in a speech. Clearly Erroll still thought it was a good idea.

I didn't quite share his view. I thought it was a fatuous description of a hard task. It was about as appropriate in its setting as the 'wind of change' phrase in Macmillan's speech to the South African Parliament. I thought it strange that a President of the Board of Trade should be so remote from the people who were actually involved in the export business that he could provide such an irrelevant and—to the people involved—irritating argument.

I commented on this in November 1963, when I addressed the conference of the Institute of Directors, most of whose members were shortening their lives and getting ulcers flogging around the world selling in the face of fierce international competition. I had just been stumping Europe for export business and my comment, which many of the directors attending the Albert Hall meeting apparently shared, was: 'I personally find exporting intensely difficult and tiring; interesting—yes; essential—yes; but fun—decidedly no.'

But to return to Erroll. To my way of thinking he made no contribution to the forward thinking about energy that was so essential. Perhaps he had decided then that his political star was descending because in 1964 he was created a Baron and, apart from a few infrequent speeches in the Lords, has since almost completely disappeared from politics. Two years after he left the Commons he was appointed Deputy Chairman of the Decimal Currency Board. So the first of the many changes in Ministers that I was to experience made little impact upon me.

Some of the Ministers I had to deal with were frankly inadequate, others were good and some mediocre. One thing they had in common though—not one of them was apparently able to think boldly beyond the advice given by their civil servants. Harold Wilson and George Brown soon found that the fine ideas they had when they were in Opposition had to be fathered by inexperienced and, in some cases, weak Ministers, incapable of standing up to their civil service advisers. Departmental Ministers were all too often left alone in their sea of inexperience, and too little attention was paid to them by men with previous experience of office. I always had the impression that Harold Wilson, having made a speech about a problem, thought something had actually been accomplished. So, when he appointed Ministers I had the feeling that he didn't follow through with the necessary accountability until the shortcomings became obvious. I was able to see the impact of political decisions at close quarters through my membership of the National Economic Development Council since its inception early in 1962.

The Chancellor, Selwyn Lloyd, was responsible for setting up this organisation and described it as 'a major step in Britain's economic history'. He was quite right. It is the only forum where the Government, the trade unions and the bosses can meet together to discuss economic problems and have the opportunity of doing some forward thinking. Neddy has done some very valuable work and it could have been still more effective; it has certainly justified the time given to it by Ministers and the TUC and employers' representatives.

When I retired from the NEDC in June 1971, I left behind the sole remaining survivor of the original team that had been brought together by Selwyn Lloyd in 1962 and that was Sid Greene, now Sir Sidney, the railwaymen's leader. Membership of the NEDC was one of the most fascinating assignments that I had. It was marvellous to watch the procession of Ministers that came and went, note their style of presentation and the impact they made upon the hard-headed TUC and business representatives.

George Brown in my view made a grave mistake when, with the setting up of the short-lived Department of Economic Affairs, he virtually wiped out the NEDC's effectiveness. Had he used the Council for the production of the famous National Plan, it would not have been the disaster that it proved to be.

The 1964 election had brought Wilson to Number 10—his

life's ambition realised—and the miners were overjoyed. Now, they thought, the coal industry would be all right, for at the helm were the men who had promised the miners so much. That they were to be bitterly disappointed, even to the extent of some miners withdrawing their political affiliation fees, was in the future. For the moment there was a splendid state of euphoria.

There was every reason for the miners' optimism, for the Labour Party and the TUC had in September 1960 (a couple of weeks before I joined the Coal Board) issued a Fuel Policy Statement prepared by a team headed by Harold Wilson. Among the proposals contained in this Policy were that there should be a tax on fuel oil, that some oil-burning power stations should be converted back to coal, and that there should be regulation of oil imports to help the balance of payments. It was strongly argued that it would be folly to let the native coal industry run down, thus placing industry and domestic users increasingly at the mercy of oil imported from politically unstable areas.

At the fairly stormy Labour Party Conference in 1960, Wilson summed up the debate on Labour's Fuel Policy, closing with the words:* 'Comrades, the mineworkers have always stood at our side even in the darkest days, so let us show by passing this resolution, that we are standing by them now.'

In the House of Commons on 7 November 1961 he said:† 'Three years ago we called for a national fuel policy, for a figure that the Government would honour for the size of the industry. Two hundred million tons was mentioned as a figure for the national indigenous coal industry to work to. This would have meant controlling fuel oil imports and controlling other things as well, but the Government insisted on what they called freedom of choice—by which they meant, of course, a refusal to touch the profits of the oil companies.'

The miners thought this was good stuff, but of course they were not to know what was to come.

The Labour Party went further in the run up to the 1964 General Election and made some very clear and sensible promises about the future of the coal industry. George Brown announced that one of the earliest acts by the new Government would be to set a figure on the future size of the industry. He made no

* 1960 Conference Report, p. 207.
† Hansard, 7 November 1961, column 921.

bones about saying that in his view that figure would be 200 m. tons of coal a year.

When Labour came to power and was in a position to do something about the coal industry, however, these promises were soon forgotten. The tragedy is, of course, that the judgments Labour formed in Opposition were perfectly correct. If they had stood by their policy in office, the country would not in 1971 have had to scour the world for imports of coal, paying prices far higher than the British prices, and having to cut back on exports. It would not have been necessary to add heavily to the burden of balance of payments by converting power stations from coal to oil.

All this would have been avoided and we should have been able to increase vastly our exports to the Continent, where the other coal-producing countries have made precisely the same mistakes we have and have run down their own mining industries too rapidly. Furthermore, the Labour Party would have benefited politically because it would not have been necessary to close so many pits and pay out vast sums of money in redundancy compensation and Social Security benefits, not to speak of the fantastic costs of providing new jobs to take the place of old. When we closed a pit employing say 2,000 men, those jobs were lost for ever in the area, and the cost of introducing new industry was so expensive, in fact, that realistic figures have never been published.

Tom Fraser was Labour's Shadow Minister of Power just before the 1964 General Election. In June—only four months before the Labour Party came into office—Tom pledged the Party to the 200 m. tons a year. These were his words in an interview published in the *Financial Times*: 'It is essential that coal should continue to supply about 200 m. tons a year of the UK's total energy requirements, and this should be set as a target for the next four or five years.'

Surely nothing could have been clearer than that, especially as George Brown had given a similar undertaking in writing to the NUM. But when Harold Wilson announced the names of his Ministers, Fraser (an ex-miner) had been made Minister of Transport and the Power Ministry had been given to Fred Lee. Further to the left of his Party than Fraser, Lee was constantly pressed about Labour's pre-election pledge and Transport House finally adopted the simple expedient of declaring it had never been made.

The symptoms of the Labour Government's later failure were quickly apparent in the way they crumpled under the advice of the civil servants and abandoned this promise. It was sheer lack of the courage and determination needed to fulfil the judgments and policies they had formulated in Opposition. In this respect, as in many others, they were a push-over for the conventional wisdom of their civil service advisers.

The NUM leaders finally had to acknowledge in July 1965 that the Labour Government did not intend to stand by their undertaking to maintain coal production at 200 m. tons. In his presidential address to the NUM's annual conference Sid Ford, who must have felt deeply disappointed, quoted statement after statement made by Harold Wilson, George Brown and Ray Gunter over the preceding eight years making commitments to the coal industry. But in the end he was forced to say that the Government were not now giving any such firm undertaking. He rubbed in the difference between the Labour leaders' attitudes in Opposition and in Government with this comment: 'The failure of the Government to take urgent action along the lines of the policy which has been consistently advocated by the Labour Party in Opposition has been, in my view, an unfortunate and grave omission.'

Two days later Fred Lee addressed the conference as Minister of Power. He had always been one of the favourites of the Left but he must have felt his lack of popularity with the lads on that occasion. Not surprisingly, his attempts to reassure his audience convinced nobody. The atmosphere remained one of disillusionment and almost incredulity that the Government should be so unfeeling and, indeed, undependable. When I gave my traditional conference address two days later I had to try to get the industry to face the harsh reality of the situation. I said that I could not understand why coal should, in an expanding energy market, be singled out for a declining role, but acknowledged that we would have to get our costs down to keep the industry as big as we possibly could. 'The brutal fact is,' I said, 'that our prices are too high, and in order to sell 200 m. tons we have got to get our costs down.' This could be done by a redeployment of labour within the industry, high productivity and a reduction in unofficial stoppages and absenteeism. The previous year the industry had lost from these causes 1·3 m. tons of coal worth about £3 m.

Absenteeism had added another £3·5 m. to our costs. I made it clear that the slackers were a minority and that most men put in a full week's work. However the big problem was that there was a violent increase in absenteeism on Mondays and Fridays which made it difficult to man the pit efficiently.

I have no doubt that most delegates went away from the Margate conference that year feeling that it was a very hard world.

Fred Lee's appointment as Minister of Power came to me as a great surprise. He had been my Parliamentary Secretary when I was Minister of Labour in 1951 and we were young men together in the Manchester area, working in the trade union and Labour Movement. I knew Fred well and realised that he was to the left of the Party; I had always been to the right. He belonged to the 'Keep Left' group in the Parliamentary Labour Party and I was a leading member of the Gaitskell group. Some of Harold Wilson's ministerial appointments were suspect, and I felt his appointment of Fred Lee showed how little consideration he had given either to Fred's ability (which lay in the completely different fields of labour relations), or the task which he was being invited to perform. Or was he just filling jobs on the basis of ensuring the balance of power between the right and the left as all Prime Ministers have to do. It was my view at the time that, as a result of this, some inappropriate appointments were made. Certainly if the best use was to be made of Fred Lee then undoubtedly he should have been at the Ministry of Labour or with George Brown at the DEA. Here he could have concentrated his whole attention on the impact of the National Plan upon workpeople and upon ensuring that the lines of communication were maintained to factories and workshops throughout the country, so that the rank-and-file workpeople could have been prepared for the traumatic experience that 'dragging Britain screaming into the twenty-first century' would bring. During the war he was the convenor of shop stewards at Metropolitan Vickers and knew every trick of the trade. Few people have had Fred's experience in dealing with thousands of workers and their problems during war time. We both entered the House of Commons in 1945.

His appointment turned out to be a complete mistake. He never had a chance, for his background knowledge, experience and training had not equipped him to be the head of an economic ministry. Harold Wilson had made an error. Before the Labour Government finished their six years of office, Fred Lee was a backbencher once again. It wasn't his fault either; he was a square peg put into a round hole by a man who should have known better.

Fred just did not match up to his big job, with coal, gas, electricity and oil to handle, and with steel nationalisation to follow. He was very unhappy and I think that on the quiet he knew he was out of his depth. The civil servants at the Ministry of Power were too strong for him, though he certainly could have stood up to the officials in the Ministry of Labour, who would have found him a knowledgeable man in the subjects he had grown up in. Eventually Harold Wilson removed him from the job, but his period as Minister gave me a series of headaches.

As I have said, George Brown had given his undertaking to the miners that the Labour Government would fix a size for the industry and that his opinion was that the figure would be 200 m. tons. Harold Wilson had also, in Opposition, encouraged the miners to expect this.

It was not long before the hedging on this pledge started. Within a few weeks of the Election, Fred saw the miners' leaders and, although he seemed to have given them the impression that the Government would carry out their election pledges, he did not actually commit himself to the 200 m. tons a year. Embarrassment over what the NUM certainly regarded as a clear promise may, indeed, have been the reason why Harold Wilson chose him for the Minister of Power. A good left-wing trade unionist would obviously be better able to sell to the left-wing of the Party and of the miners a figure lower than the 200 m. tons. Wilson would know that the right-wing would always behave itself whilst a Labour Government was in office. Early in the following January there was evidence that the Department had already been working hard on poor Fred. He saw another NUM delegation and again avoided committing himself to the 200 m. tons. Instead phrases like the 'base load' started to be used to describe coal's share, with other euphemisms like 'a very substantial part of the nation's fuel needs'. And the miners' leaders were told that the Government were considering the possibility of a National Fuel Policy—which the Union were known to want.

But of course, they were not to know that the Fuel Policy announced by another Labour Minister would propose that the output should be down to 120 m. tons in 1975 and 80 m. tons by 1980. Which meant that the manpower was to be reduced to 165,000 in 1975 and 69,000 in 1980, a total reduction of 312,000 men from the number then employed. On my own insistence, in fact, the latter figure was taken out of the White Paper that was eventually published in 1967. But Dick Marsh, Fred Lee's successor, forgot and let the figure slip out in the House of Commons. But that all came later. In 1965 I sensed what was happening, of course, and took an early opportunity to say in public that I considered the Government were pledged to the same aim that I had—sales of 200 m. tons a year. I took a lot of trouble to spell out how the total had been derived, demonstrated that it was not a nice round figure plucked out of the air, and warned that if the target were lowered, even by 10 m. tons, the Coal Board would have to add £20 m. to prices. And it is exactly that which has pushed us into price increases in recent years after a long spell of price stability.

Poor Fred must have been having a terrible time of it in those first months in his job, because a couple of days after I had spoken the NUM released the text of the letter George Brown had sent them in May 1964, in his capacity as Chairman of the Labour Party Home Affairs Committee. This stated quite clearly that the Labour Party estimated that the industry's output must be maintained, at least in the short run, at around 200 m. tons a year, with the prospect of extension should this become necessary. George's letter went on: 'We accept that if the coal industry is to perform its role efficiently, its output must not be subjected to violent fluctuations and stop-go policies. There must be, therefore, some reasonable guarantee to the industry that its target output will be absorbed. To this end, the public sector as a whole, which now takes roughly half the industry's output, can make a vital contribution. In addition, temporary fluctuations in demand for coal from the private sector can be evened out by stock-piling measures to the cost of which the Labour Government would be prepared to contribute.'

Such good sense, but so quickly forgotten under pressure from a few civil servants!

Even Sid Ford, President of the NUM and deeply loyal to the

Labour Party, had again to criticise Fred Lee's evasions. Sid argued that Lee should have gone into his Ministry and said: 'I am here to carry out Labour Party policy. We are going to have a national fuel policy and it is going to be based on 200 m. tons of coal.' However Transport House said that they knew of no party commitment to 200 m. tons. The last official statement on fuel policy by the Labour Party had contained no figure and there was nothing said about it in the election manifesto.

So the miners naturally enough went to see the man who had made the promise to them, George Brown, then Minister for Economic Affairs. George was quite prepared to stand by the Government's pre-election pledge. I was present at this meeting with the NUM leaders, and so was Fred Lee. I wondered what Fred, having been thoroughly briefed by his officials, thought of his Deputy Leader's promise. Anyway, the NUM said on leaving the Ministry that they were a lot happier as a result of the meeting.

Two days later, Fred Lee told the Commons: 'The Government accepts for the present the case for trying to maintain the position of coal (deep-mined and opencast) at around the present level of 190 to 200 m. tons.' He added that ways of assisting the industry's efforts accordingly were being studied. The words he used had obviously been very carefully chosen, but the planning for a much lower level was still going on in his Department.

Within a month of Lee taking up his job as Minister of Power, the Chairmen of the Nationalised Fuel Industries were called to his office to begin the foundations of what was intended to be a co-ordinated fuel policy. This body in effect was the revival of a group known as the Minister of Power's Co-ordinating Committee, which had not met for several years. In the event, it would have been just as well if it had never been revived. It turned out to be no more than a talking-shop. No decisions were made by the Committee and the Minister read his brief—every word of it. I have never been involved in a greater waste of highly paid people's time. But it put the seal on the official document. We had been consulted.

In the spring of 1965 I got thoroughly fed up with people arguing that the coal industry was being protected and feather-bedded. In public speeches and through *Coal News* I pointed out what Exchequer money had gone into other industries. We did not, of course, question the wisdom of these arrangements, but merely contrasted the generous treatment that others had had compared with the coal industry which, since nationalisation in 1947, had not cost the British taxpayer a penny, although it had been quite substantially subsidised before nationalisation. The biggest beneficiary, of course, had been agriculture, which had received about £2,700 m. in subsidies and grants over a ten-year period. The white fish industry was being helped in 1965 to the tune of £4½ m. in a year, and the Forestry Commission was handing over £2 m. that year to private growers and developers. The Government had written off £110 m. of the accumulated deficit of BOAC. British Railways had enjoyed the writing-off of part of their accumulated deficit to the tune of £475 m., while a further £705 m. was put into suspense account. I knew that our own capital reconstruction was imminent and, when it came, I didn't want anyone repeating the accusations of feather-bedding.

I had advised the then Minister, Richard Wood, during my first year as Chairman, that a capital reconstruction of the industry would be necessary since many of the original assets had long ceased to exist and ought to be written off. The Minister did not disagree with this advice, but we decided to see the future pattern of the industry before proceeding with detailed calculations.

However when the financial results for the six months ending September 1964 were published, they showed a deficit after interest of £18 m. In fact, there had been an operating profit of £2 m., but interest payable to the Minister was £20·5 m. I thought it time to raise the question of capital reconstruction again. Furthermore, the Coal Board had invested new capital to provide for an output of 240 m. tons of coal a year. This was under an earlier policy decision taken in 1950: even at this proposed level the then Federation of British Industry regarded it as too small and urged a planned output of 270 m. tons.

A good deal of this capital in the result was abortive, but had to be serviced.

When in March 1965 it was announced that BOAC's debts of £110 million were to be written off, it so happened that I was addressing a private meeting of the miners' MPs at the House of Commons the following day, and I took the obvious opportunity to argue the coal industry's case for similar treatment. I pointed out that BOAC was a comparatively new industry, but that this had not prevented the Government from accepting the case for a substantial write-off. Of course, the Minister responsible in this case was not Fred Lee. One of the reasons for BOAC receiving this treatment was the set-back caused by the Comet failure. But we had suffered from an equally serious catastrophe in the loss of millions of tons of business to the nuclear power stations without a scrap of commercial justification.

The capital reconstruction came in July 1965, about four years after I had advised the then Minister that it would be justified. The Government wrote off £415 m. of our capital debt. The justification for this was not hard to see. It was not that inefficient planning of the pits by my predecessors was responsible, or bad management of the industry, but the over-estimation of coal requirements. Pits are not sunk in a day and millions of pounds had been invested in new collieries and in major reconstruction of existing pits. Among these schemes there were some very risky mining ventures which neither the mining engineers nor the geologists would have selected if it had not been for the extreme pressure to increase the industry's total capacity. At that point of time when the decisions were made, it could not be said that they were the wrong decisions to take. No one had foreseen the vast quantities of cheap oil that became available seven or eight years later—temporarily, as it turned out.

By the late 1950s it had become clear that in fact the capacity was no longer going to be required and the 240 m. tons objective was quietly forgotten. But, of course, the Coal Board were left with the debt on the investment, some of which was in schemes which were still only partly completed, but from which it was much too late to draw back. Older pits had to be closed more rapidly than they would otherwise have been because too much capacity, combined with the market contraction of the time, left pits insufficiently depreciated.

Fred Lee never seemed really to understand this, and in his speeches gave the impression (and I am sure he genuinely believed it) that somehow or another he was making some kind of gift to the miners. It was nothing of the kind. The gift was to coal consumers, notably the bigger ones like the electricity industry, whose coal prices were relieved by the Board having to extract from them about £21½ m. less a year for interest. Lee said that the capital reconstruction had been agreed because of the contraction of the industry, and because of future pit closures.

The truth was that with the contraction of the industry the value of the assets was no longer there. Furthermore the liabilities included the loss of £70 m. on imported coal. The NCB, at Government request, had imported large quantities of coal at high prices and sold it at the much lower inland prices. There was also a £25 m. loss as a result of the Government stop on a price increase, which we needed because our costs had gone up with inflation, in defence of the prices and incomes policy which George Brown at that time was struggling hard to secure. Since that date the pit closures have gone on apace and in 1975, in my view, there will require to be another valuation of the fixed assets, which may well reveal that a further capital reconstruction is necessary.

The row that developed over Labour's pre-election promises was given another twist when in January 1965 three Ministers to whom I had appealed to stop conversions from coal to oil in hospitals, army camps and one of Britain's largest ordnance factories rejected coal. The Ministers involved were Kenneth Robinson (Health), Fred Mulley (Army), and Charles Pannell (Works). Naturally, the rejection was taken by the NUM, as well as by the Coal Board, as an ominous sign of the new Government's intentions. Sid Ford came out with a statement in which he said that the Union acknowledged that the Government could not tell domestic consumers that they ought to use coal instead of oil, but that the Union did expect the Government to favour our own indigenous fuel in establishments which they controlled—a not unreasonable argument, one would have thought.

Came the General Election of 1966 and came a new Minister of Power—Dick Marsh—my fourth in just over five years. In our first dealings Marsh was obviously uncertain about me and appeared ill at ease. However he soon realised that I had a high regard for his ability and that I was ready to co-operate with someone who was prepared to do his own thinking, as he obviously was. This made a big change for me after having to listen for two years while official briefs were read to me. I know he was a good Minister but I gave him a lot of trouble because he was too ready to close pits, a bit over-eager to usher in the white-hot technological revolution. He was going to go down in history as the Minister who brought in the North Sea gas.

Our most vigorous tussles were over the 1967 White Paper, and I have no doubt that now, with hindsight, he regrets not following my advice and arguments in 1962. If he had done so there would have been no need for expensive coal imports in 1970/71. His errors were the mistakes of inexperience, but I think Harold Wilson again blundered in moving him to the Ministry of Transport and out of the Cabinet. Dick Marsh is now Chairman of British Railways: a great loss to the Parliamentary Labour Party, a boon to Railways.

When a Labour Government is in power it is the NUM's custom to invite the Minister responsible for the coal industry to speak at their annual conference. Richard Marsh had this experience during his time as Minister of Power.

Although Marsh bluntly told the miners about the progress being made by other fuels, he thought to encourage them by denying reports that deep-mined coal production was to be cut back from its then level of 167 m. tons a year to 140 m. tons by 1970. He spoke of a figure for that year 'substantially in excess of 140 m.'. But he turned out to be wrong. Deep-mined output did come down to 139 m. tons in 1970. I am not suggesting that he deliberately misled the miners. But his forecast, sincerely given at the time, nevertheless did prove to be over-optimistic.

I told the annual conference of the NUM at Eastbourne in the summer of 1967, that Dick Marsh was making a genuine attempt to balance the competing claims of rival fuels and to shape a policy which would provide the nation with cheap energy in the long run. I said that his decision to hold a conference at the Selsdon Park Hotel a couple of months before the NUM met was something of a milestone because it was the first time that the nationalised fuel industries had got together with the Minister and his officials to thrash out policy problems. I paid tribute to Dick Marsh for this imaginative step but whilst doing so I pointed out that I might not necessarily agree with his conclusions which were expected to be announced that autumn. In particular, I certainly could not accept some of the more pessimistic figures for coal output which had been quoted in the Press.

One of Dick Marsh's moves was to appoint Cecil King, Chairman of the International Publishing Corporation, as a part-time member of the National Coal Board in July 1966. Marsh once said that he had chosen the strong-willed King to act as a brake 'on the enthusiastic bulldozing of the Chairman of the Coal Board'. However he found that the man he had appointed as a brake quickly acquired the enthusiasm that the rest of us already had and, far from acting as a brake, further motive power had been provided. King was as puzzled and outraged as anybody at the antics of the Labour Government by the time he finished his three-year term of appointment.

It is only fair to Marsh to say that before appointing Cecil King he consulted me, as is the custom, and I indicated my pleasure at the proposal. Only one Minister ever made an appointment to the Coal Board without consultation and that was Roy Mason. A breach of etiquette.

In his memoirs *Strictly Personal* King recorded that Marsh was deeply disappointed to find that although he had appointed King as a counterweight to me, in fact he became my ally. In his book he wrote:* 'I would not run the meetings of the Coal Board as Lord Robens does, but that is irrelevant. Every Chairman must run the business for which he is responsible in the way suited to his temperament, and he is entitled to be judged by the results.' He was kind enough to add: 'By this standard Lord Robens emerges with all colours flying.'

While he was a part-time member of the National Coal Board, King was at the centre of an odd incident involving the miners' Members of Parliament.

I used to arrange for the MPs to come to Coal Board Headquarters once a year for a meeting at which I would make a short statement about the industry's prospects and they would ask questions. As an ex-MP I was always a great believer in the value of contacts like these. Naturally, the part-timers, as members of the Board, were informed of these arrangements and were welcome to attend, and entitled to be present.

In May 1968 King had publicly attacked Harold Wilson's leadership of the Government and his handling of the economy. In fact he had the temerity to suggest that Harold must go. When the MPs arrived and found that King was expected to come to the

* Cecil King, *Strictly Personal*. London: Weidenfeld & Nicolson, 1969.

meeting, they sent a message to me saying that they would refuse to sit in the same room. Unless I asked King to stay away, they would walk out.

Joining them in the conference room well known to generations of NUM negotiators as Room 16 a few minutes before the meeting, I found that none had taken their seats. Board members and MPs were standing around. I talked to the miners' MPs and certainly not all of them were for refusing to sit at the same conference table as Cecil King. I made it clear that under no circumstances would I ask a member of the Board to withdraw. They had asked to meet the Board and King was a member. If they did not wish to meet all the Board members, there was nothing I could do about it. They were asking for something that was just not negotiable. If they did not want to go ahead with the meeting to discuss the problems of their constituencies, and their members who worked in the pits, I regretted it. But I certainly had no intention of asking King to leave.

When this was made clear the MPs left and went straight back to Westminster, where they tabled a motion asking Ray Gunter, then in the middle of his short stint as Minister of Power, to remove King from the Board. This motion was signed by no fewer than 155 Labour MPs, but King was allowed to finish his three-year appointment which expired the following year. He was not re-appointed. In Britain we don't shoot or imprison people we don't like or disagree with on political matters, but there are other methods of showing displeasure.

Giving evidence to the House of Commons Select Committee on the Nationalised Industries in June 1967, I complained of delay on the part of the Ministry of Power in clearing capital investment schemes put to them by us. Richard Marsh went to the Committee for his turn the following week and argued that his Department was very much a scientific engineering ministry with very few scientists. In fact, apart from qualified people working in the Safety Inspectorate, there were virtually no mining engineers or scientists in the whole of the Ministry of Power.

No doubt the Labour Government thought that when the White Paper was published in November 1967 the public debate about the relative costs of nuclear power and coal would come to an end. On the contrary, I continued to clamour for an independent inquiry into these matters, which were of vital importance to the future of the coal industry. In public statements Richard Marsh repeatedly said that coal must be competitive. I always asked the simple question: 'Competitive with what? We know the price of oil, we know the price of gas, we know our own prices. What we don't know are the real costs of nuclear power.' In January of the following year the pressure on me to close my trap and cease embarrassing the Labour Government was terrific. At Westminster the anti-coal lobby (that is, the nuclear power pressure-group) were whispering in the ears of Ministers that they should not stand for such behaviour. Chairmen of nationalised industries were suppposed to take instructions, not to argue the toss.

Naturally, I took a different view. Parliament had laid down in the statute nationalising the coal industry the duties and obligations of the Coal Board with great precision. At least I knew sufficient about the Parliamentary game to realise that not even Prime Ministers can contravene Acts of Parliament. So I went on with my demands for the information we must have if we were, as the Members of the National Coal Board, to carry out our public duties. I used to say: 'If someone tells me accurately the costs of nuclear power, I then know whether there is any chance of competing. If it is quite clear that we have no hope, then we can all start taking life more easily, put on our jackets at 5 o'clock every evening and clear off home. But if there is any hope that we can match the real price of nuclear-generated electricity, then we will go on fighting to carry out our statutory duty to supply coal in quantities and at prices to meet the demand.'

Marsh tried to damp down the public argument by saying, in a speech to the Coal Industry Society in February 1968, that there could only be one fuel policy and that was the responsibility of the Government. He got his policy, but it turned out to be nothing to be proud of.

In this speech, with his characteristic wit, Marsh rebuked the heads of the nationalised fuel industries for their public wrangling. He said: 'We are a sporting nation and sometimes the rougher the game the more spectators like it. We have a special liking for the type of wrestling which takes place on television every Saturday afternoon. I sometimes wonder whether next Saturday I shall be seeing Alf Robens versus Stanley Brown and Ronnie

Edwards versus Henry Jones, with the referee being clobbered by all four.' Clearly Marsh regarded himself as the referee.

When I addressed the same audience just a month later, I continued the Minister's wrestling analogy. I pointed out that a wrestler was entitled to two public warnings before he was disqualified. I went on to say that I thought perhaps I had received my first during the previous month and added: 'It may be that after today I shall receive my second.'

I made it absolutely clear that I should continue to speak out on matters that were certainly of the greatest importance to the mining industry, and were also of great public significance. Just for good measure, in a talk to the miners' MPs the next day, I attacked the Prime Minister and George Brown for breaking the promises they had made to the miners in 1964. I also said that the planned rundown for the coal industry was too fast for the country to bear. I pointed out the folly of the Government spending enormous sums on nuclear power stations although they had no clear idea of what the electricity they would produce would really cost. Finally, I said that hints that I might be sacked if I did not cease my attacks would not deter me.

By one of those strange ironies in timing, I attended a dinner at Number 10 Downing Street that same evening. I got the feeling that night that I was not my former colleague's favourite guest.

Soon after Harold Wilson's visit to the Durham Miners' Gala in July 1967, when he came under such pressure to make the Seaton Carew power station in the county coal-fired (see the next chapter), he was obviously thinking also of the shocks to the miners that were going to come when the Fuel Policy White Paper was published later in the year, and he announced that the Coal Board was being ordered temporarily to stop pit closures. It was said that initially the standstill would last for one month— that is until the end of September of that year. This would save the jobs of about 5,000 men at five collieries. The Government were going to meet the cost which could, it was estimated, be as much as £500,000 for the one month's delay.

The miners' leaders were gathering for the TUC conference at Brighton and Will Paynter made it clear that he, for one, was not taken in by this manœuvre. He accused the Government of 'a shabby political pantomime' and described the announcement as 'nothing but window dressing'. Paynter also said: 'It is a scandal and a phoney.'

Although the decision was only going to pile up a greater number of pit closures later on, I was moderately pleased about it, because if we had announced closures in September it would have meant a very austere Christmas for a lot of people, and we had made a practice of trying to avoid pit closures during that season.

In fact, the Government announcement did not do the Prime Minister much good at the Trade Union Congress. In spite of his postponement of the closures, the NUM delegation voted in favour of backing a conference resolution censuring the Prime Minister over unemployment.

Wilson was at last learning that the miners could not be automatically relied upon for dog-like loyalty. Clearly they were angry with the Labour Government and they were annoyed with him personally. Indeed, miners' leaders at that time said to me that a Tory Government would never have pushed the miners about as a Labour Government did. They thought that Wilson was taking for granted that they would be blindly loyal. And, of course, nobody enjoys being taken for granted.

A couple of weeks after the Trade Union Congress, Bill Webber, the Industrial Relations Board Member, and I had a meeting with Wilson and Dick Marsh, his Minister of Power, at Downing Street. The Government were under pressure to continue the temporary postponement of pit closures and to treat more pits in the same way. Dick Marsh was very obstinate at this meeting and kept insisting that any deferred closures must be added to those in the following year. The PM was also firm and at one point admonished Marsh in front of Webber and me. I had to say very firmly that the miners had taken a great deal of punishment and it was clear that they were at breaking point over pit closures. The pace was far too fast and only the miners' intense loyalty to the Labour Government and the great care we were taking at considerable cost to mitigate the hardships and the difficult social consequences had so far prevented industrial disputes.

Finally I said that the Government were in danger of planning themselves out of office. It was a grimly prophetic statement.

Wilson made his announcement about further deferments to

the miners' leaders assembling at Scarborough for the Labour Party Conference at the beginning of October. He told them that sixteen collieries, due to be closed in the following three months, would not after all die before the end of the year, at the earliest.

If he expected the miners' leaders to be grateful for this compassion, he must have been painfully disappointed. Will Paynter said after the encounter that it had been an angry meeting, although it was difficult to remain angry with someone who stayed as imperturbable as the Prime Minister. Paynter added however that the current rate of closures was causing cynicism and demoralisation in the mining industry.

Wilson no doubt had a number of motives for his actions. Perhaps he was trying to soften the blow that he knew was coming quickly in the form of the Fuel Policy White Paper. But he was only tinkering with the whole problem and the miners knew it.

Deferring the closure of the sixteen pits was a device the Government never adopted again. It added an extra burden to the programme for the following year so that we had to cope— the unions, the men and the Board—with the highest number of closures ever known in a single year. Again, the Prime Minister was tinkering to try to limit the impact of a basic policy which was ill-conceived and worked out by the civil servants. In Opposition, Labour had the right policy for the coal industry; in office, it ran away from it because the civil servants were too persuasive and determined for the Ministers Harold Wilson appointed.

Wilson, when Prime Minister, had an extremely unhappy visit to South Wales in February 1968 at a time when we were about to close two collieries, Cefn Coed and Yniscedwyn. The night before his visit the Minister of Power had issued a statement saying that the Coal Board had not followed the proper procedure and therefore had been wrong to announce that these pits would close the following month and that the Board intended to serve the notices a few days after the Prime Minister's visit. While he was in Wales the Prime Minister backed up the Minister's statement.

Until we had announced a date for closure, the local Economic Planning Council could not decide whether to ask for postponement.

But we had consulted the NUM on both pits, each of which had been given a trial period. The NUM leaders in South Wales were not taken in by the Prime Minister's window-dressing. Glyn Williams, the President of the South Wales NUM, said that if there was to be a deferment, the Government would have to foot the bill. The Secretary of the South Wales Area, Dai Francis, said the trouble was that: 'Lord Robens is in a political battle with the Government and Mr Marsh, the Minister of Power.'

The Prime Minister's car passed through Cefn Coed and, to the amazement of the local people and television crews and reporters, drove straight through—without stopping. All the miners could do was to wave placards bearing slogans like 'Coal Not Dole' and 'Please Let Us Keep Our Jobs' at his car as it swept past. However he did stop later to speak to NUM representatives from the two pits.

Between 1965/66 and 1968/69, we closed no fewer than 204 collieries. This rate of a pit closure almost every week for four years was achieved under a Labour Government. In fact under Labour, perhaps because of Harold Wilson's enthusiasm for technological change, we shut pits at a faster rate than when the Conservatives were in office.

Although the closure programme was bigger than we wanted it to be and has proved to be bigger than it ought to have been, we were never in any doubts where we were going. In July 1966 I told the annual conference of the National Union of Mineworkers, meeting in Scarborough that year, that in five years' time (in other words, in 1971) pit closures, except where reserves were exhausted, would be a thing of the past.

I also said that by the end of 1970 we would be left with about 300 pits that could produce up to 180 m. tons of coal a year but that they would have an output of only 130 m. tons if the manpower drain continued at the rate we were then experiencing. Well, the run-down did go on for several years after 1966 and, by the time 1970 came around, we had about 290 pits which produced together 139 m. tons of coal.

The idea that Britain should shed most of its coal industry, with all its attendant problems, was widespread in the Civil Service five or six years ago. I remember when James Callaghan was Chancellor of the Exchequer and I had been trying to persuade him of the good sense in maintaining a healthy indigenous fuel policy. But he said to me: 'Well, Alf, I'm afraid I couldn't find an official in the whole of the Treasury who would accept your argument.' Presumably those Treasury officials have now amended their views. Once again, this country has had to learn a lesson the hard way.

By the spring of 1968 the Labour Government was already in desperate straits. The row over the compulsory incomes policy was beginning to tear the Party to pieces. In a bid to save the Government from disaster, Harold Wilson carried through a Cabinet reshuffle, appointing Barbara Castle to the hot seat of Secretary of State for Employment and Productivity. Mrs Castle displaced Ray Gunter as Minister of Labour and Ray was appointed Minister of Power in place of Dick Marsh, who moved to the Ministry of Transport.

I am not letting any secrets out when I say that Ray, an old Parliamentary friend of mine, was not too pleased about the move. So again, I found myself with a Minister who did not want the job. Like Freddie Erroll in the Conservative Government, he thought he had been down-graded. Ray stayed less than three months and resigned because of his dissatisfaction with the way Harold Wilson ran the Government. He made no impact on our problems, so I am not being disloyal to an old and respected colleague when I say that his departure, like his arrival, did not make a ha'p'orth of difference to the coal industry.

Gunter's successor as Minister of Power was the former Yorkshire miner, Roy Mason. Before he had been a month in office, he told the miners' MPs that he had begun a complete review of the Government's Energy Policy. It soon turned out that this review was going to have little in it to comfort or encourage the mining industry.

During the Parliamentary summer recess, Mason announced what was perhaps the biggest single blow to the future of the industry. This was when he allowed the power station at Seaton Carew to be nuclear powered. The full story comes in the next chapter.

I did not doubt Mason's sincerity in wanting to do his best for the industry. He was just another Minister who came into office with the right basic ideas and then allowed his judgment to be undermined by the civil service advisers. He was not alone in this, of course. When senior members of his Party like the Prime

Minister and the Deputy Leader allowed themselves to be pushed off a sensible course for the coal industry, it would be unreasonable to blame him particularly.

While he was Minister of Power Roy Mason gave evidence to the Select Committee on the Nationalised Industries but was smartly criticised for his failure to speak up for the coal industry on that occasion. The Derbyshire Area Council of the NUM, representing nearly 20,000 miners, passed a resolution of no confidence in his ability to give the leadership the industry needed. What apparently particularly angered them was that Mason was an ex-miner and they thought he should have had a better understanding of the needs of the industry. Nor was dissatisfaction with the miners' own Minister absent from his own constituency. There was talk at one time of nominating another candidate to oppose him. In Barnsley of all places!

Our pressure upon him to have an inquiry into nuclear power costs in public or private was continuous. We knew that we were being fobbed off and that in many places coal could compete, but not against the artificially low costs that we were being fed. Frankly we regarded them as phoney, as indeed they turned out to be. Meanwhile the coal industry was being crucified.

Mason had an extraordinary reason for rejecting all our calls for this independent inquiry—or at least he gave the Yorkshire miners one. He said he had refused the inquiry because he was satisfied that nuclear power was cheaper and that any examination would prove that and shatter the morale of the miners. He said the result of any examination would ultimately be to kill coal. He went on to say that as far as he was concerned it was the miners' morale that mattered most. What a tail our cat's got, as we used to say when I was a lad!

In my long experience I have known Ministers to utter nonsense, but this I thought was the limit. What a low opinion this ex-miner turned Minister must have had of the spirit of his former colleagues! Miners never will admit to being beaten, as he must really know in his own heart. But the crushing ineptitude of his comment lay in the suggestion that if everybody knew what he knew about the comparative costs, they would give up the struggle. He was telling them that the industry really had no hope. What a way to build up morale! Surely, a serious statement of that sort ought never to have been made until an independent inquiry had established the facts beyond all reasonable doubt.

I swiftly said that I for one wasn't afraid to know the truth and continued to claim an independent inquiry as the least the industry could decently expect. But he continued to refuse. As a Privy Councillor bound by an oath of secrecy if required, I asked if I could have the figures. All to no avail. I doubt in fact if they had the figures; I suspect they took those given by the CEGB, a deeply interested party.

We not only got a new Minister in the October 1969 Cabinet re-shuffle, we also went into a new Government department. The Ministry of Power was added to Tony Wedgwood Benn's Ministry of Technology, and under him Harold Lever, as Paymaster General, was later made responsible for the nationalised industries. The Labour Government had maintained its average of a new Minister a year, but no one in the coal industry was sorry to see the back of Roy Mason, who was chiefly remembered for his decision to build a nuclear power station right on the Durham coalfield.

With Ministers changing so frequently, the only continuity of thinking came from the civil servants. They made sure that the fuel policy remained the same.

In due course Matthew Stevenson, the Deputy Secretary of the Ministry, became the Permanent Secretary. He was in the Department in these two posts for five years, leaving in 1966 to become the Permanent Secretary of the Ministry of Local Government and Housing, an administrative post as different from his previous one as could be imagined. He retired from the service in 1970, but is now a part-time member of the British Steel Corporation. He was a Treasury man to his finger tips. An extraordinarily capable civil servant, hard working with a sharp and incisive mind, he nevertheless lacked to my mind the capacity to look beyond the figures to the human beings that they represented. Some years before the 1967 White Paper, which laid down a production of 120 m. by 1975, he had been pressing this figure upon me. This was a figure which I firmly resisted, but I would not say that in doing so I carried every one of my senior colleagues at the Coal Board with me.

He certainly seemed to me to have exceeded his powers and authority on one occasion, and if I had gone to the Prime Minister I could really have made it hot for him. At the time I almost did, but I reflected that he was doing what he did with the best intention and without malice, so I let it go.

As I have previously indicated, there is pressure every year at budget time for the duty on fuel oil to be removed; at one time I was fearful that this might be done before we were ready for it. I therefore suggested to the Minister that the tax could be removed without harm to the coal industry provided that the take of coal by the CEGB were to continue as though the tax were still on. This is quite easy for the CEGB to do: they only have to feed in the figure to the computer and the rest is automatic. The effect of that simple deception on the computer would be that we should sell as much coal to power stations as if the oil tax were on, and the CEGB would save the tax on the oil they burned.

Shortly after this, Dr Schumacher, the Board's Economic Adviser, was invited to lunch by one Ministry official; when he got to the rendezvous, which was a London club, he found Stevenson was also present.

Stevenson evidently had come to the conclusion that this idea was far too clever for me to have thought of myself and assumed that it was Schumacher who had put up this bright idea. He began to upbraid him vigorously and Stevenson was never at a loss for words. When Schumacher was able to get in a word edge-ways, he said that, although he had been involved in discussions, it really was my basic idea.

Stevenson's instant retort was: 'Then you should have stopped him.' Obviously, nobody in the Coal Board was going to put up with bullying of that kind. Schumacher continued to criticise Government policy in lectures and in articles. Obviously Stevenson thought he could tell one of my officials what he ought or ought not to do, but he genuinely believed that he was better able to advise the nationalised coal industry than many of its employees or professional advisers.

The incident illustrated the power that the nameless and face-less men who walk the Whitehall corridors possess. The new gen-eration of very senior civil servants marks quite a break in the attitudes of the old, fitting into the business world with much greater understanding. The newer people coming to the top seem to be much more inclined to consider themselves part of the whole than a superior set in charge of it.

Stevenson was quite convinced that I had no intention of ever closing any pits, so that his pressure for a lower output target and a smaller number of collieries never ceased. However he went and I stayed, much to my relief. I liked him for his frankness and his absolute integrity and honesty, but he wasn't the man to deal with human problems. When he left to go to the Ministry of Housing and Local Government, his successor was David Pitblado, an old friend who, unlike Stevenson, had a great deal of humanity in his make-up and had a much better political nose.

The Commons second-reading debate on the 1967 Coal Industry Bill, which provided for financial help to ease the social consequences of the White Paper published earlier that month, produced some odd statements in speeches by politicians who, if they ever remember their own contributions to Commons debates, must do so with some embarrassment. Mrs Margaret Thatcher, then Opposition spokesman on Fuel and Power, was worried by the size of the coal stocks. She doubted whether many of them would ever be sold. Eric Lubbock, then a Liberal MP, had done an elaborate calculation. He had worked out that if all the coal then being held in stock was placed in Hyde Park, it would form a pile 14 feet 6 inches in depth, covering the whole of that area. He argued on the basis of this strange calculation that it was a shocking waste of the taxpayers' money to suppose that those stocks should be allowed to increase still further.

Mrs Thatcher is a tax lawyer and might therefore be expected to approach a social problem rather too much in terms of the balance sheet. Lubbock during his time in the House was an almost fanatical supporter of nuclear power. If the stocks that they so much regretted had still been there in the winter of 1970/71, the country would have been saved many millions of pounds of foreign exchange spent on importing coal into Britain. However one MP who could read his speech in Hansard with satisfaction even today is Sir Gerald Nabarro. He expressed his belief that coal, properly used in power stations, was cheaper than any of the alternatives, provided the station was sited correctly in the first instance. His forecast, unlike those made by Parliamentary colleagues I have quoted, has been justified by events. In fact, like the Coal Board, Gerald has more often proved right on the subject of fuel than wrong.

The Coal Industry Act of 1967 provided that to the extent that gas and electricity used coal in preference to other fuels which would have been cheaper, the cost could be borne by the Government, up to a maximum of £45 m. There was also to be special assistance to men made redundant after reaching fifty-five years of age. I know that this was a well-intentioned measure but I certainly disliked the idea of telling a man that at fifty-five the community had no more use for him, when in fact he could be quite capable of making a contribution to society for another ten years.

The Government also provided a bigger contribution to the social cost incurred by the Board in contracting the industry. And they also arranged that, when colliery closures were deferred because of social reasons at the Government's request, the Board would be compensated.

Again, I suspected the motive behind the Act. Because the planners had made the wrong decisions, the country was involved in much unnecessary expenditure. If the Government and their advisers had had our confidence, they would have paid men to go on producing coal, even if it had to be stocked for a few years before being used, rather than pay them to stay at home.

In 1971, there was another Coal Industry Act to prolong some of the features and principles of the 1967 Act which had become time-expired. This was to me a significant measure, since it caused my departure from the industry, as I shall describe later.

9

Nuclear Scandals

The advent of nuclear power was obviously bound to make a great impact on the coal industry. A nuclear power station was an irrevocable move away from coal. Oil-fired stations could be converted to coal in an extremity, provided stocking facilities and coal-handling plant could be made available. But a station built to use nuclear energy put coal out for all time. The argument for using oil as a fuel for power stations rather than coal was based entirely on the price comparison between oil and coal. In the late fifties and early sixties the cheap oil argument won the day. Whether oil was really cheap if the total costs were calculated was never considered. What I found it quite impossible to get the Government to see was that the cost of miners' redundancy, of unemployment and the loss of the capital investments in the pits ought all to be included in the total sum.

I never believed that the Central Electricity Generating Board would consider anything other than the apparent price advantage of oil; the fact that the jobs of 10,000 miners would disappear for every 2,000 MW station that used oil and that pits with a combined capacity of 5 m. tons would be closed, with the heavy capital costs and the permanent loss of reserves, was not a matter of concern to them. Nor was the increased burden on the balance of payments. It obviously ought to have been, but it wasn't. The same excuse could not be made for the Government though, and certainly not for the Ministry of Power, whose statutory function was a co-ordinating role. A role I regret it never efficiently performed. Indeed its internal organisation, with a Division for each fuel—gas, electricity, oil, coal and nuclear—prevented it from so doing, even if the Minister had wanted to. These Divisions in the Ministry just became the mouthpieces of each of the industries.

Ministers' speeches were written on the basis that coal and oil had to compete solely on price, and little or no regard was given to any other factor. The coal industry therefore had to accept that unless coal could produce electricity cheaper than oil, then it would continue to lose out to oil—as indeed it did.

When it came to nuclear power, however, it was an entirely different story. The nuclear programme originally had two main justifications—first, that we must be in at the start of this new technological development so that we could exploit its export potential, and second, that the estimates forecast that it would be cheaper than fossil fuel.

The result of this high-level policy decision and the cost prospectus, which afterwards proved to be false, was that the whole of the loss of business fell upon the coal industry. Oil-fired stations continued to be built, the consumption of oil steadily increased from 9·2 m. tons of coal equivalent in 1960 to 21·1 m. tons of coal equivalent in 1970. The nuclear power stations in operation in 1970 produced 22·8 m. kilowatt/hrs, displacing 9·4 m. tons of coal, representing nearly 18,000 jobs in mining.

All of that nuclear-produced electricity could have come from coal-fired stations, more cheaply, and at about half the capital cost.

Coal lost that business and those men lost their jobs, not for any failure to be competitive, but because there had been the policy decision to fit nuclear power into the electricity station programme. The specious arguments that accompanied this policy decision convinced quite a number of people. Even George Tyler, the General Secretary of the British Association of Colliery Management, fell for it. As late as May 1971 he was telling a Consultative Council meeting that it wasn't Government policy that had reduced coal requirements, as the Government had said that the Board could sell every ton we produced that was competitive. How naïve can you get? Every ton of the coal displaced by nuclear power was competitive.

The battle of nuclear power versus coal began as far as I was concerned in October 1960, the month that I took up my post as Chairman Designate of the Coal Board.

It is true that one of the factors taken into account when the Government had committed itself to the nuclear programme in 1955 was a possible shortage of coal. But in 1957, when coal was already coming into surplus, the Government had announced an acceleration of the programme and an increase in its size.

In June 1960 a new White Paper had pointed out that coal had become plentiful and oil supply prospects had also improved. It was acknowledged that the cost of electricity from the first nuclear stations to be commissioned the following year would be higher than earlier estimated. The White Paper admitted something else that had not been foreseen. That was a fall in the cost of power from new conventional stations.

This White Paper still spoke of the need to supplement coal and said that although oil seemed likely to remain plentiful for many years, it would not be sensible to rely only on imported oil for this purpose. What the Government of the day was really saying was that although nuclear power was uneconomic and non-competitive, the programme would be continued with in order to reduce the country's dependence on imported oil.

The Central Electricity Generating Board, which was to have the responsibility of operating the stations, had not been consulted about the size of the first programme and said in 1962 that it would not commit itself to ordering even the one station a year provided for in the 1960 White Paper.

In the very month I joined the Coal Board, Sir Christopher Hinton, then Chairman of the Central Electricity Generating Board, forecast that by the early 1970s nuclear power would be cheaper than conventional power for base load operation. The breakeven date, he expected, would be between 1966 and 1970. Well, we are now in the early 1970s and there isn't in Britain a single nuclear power station that can generate electricity within 25 per cent of the costs of the most efficient coal-fired stations.

Among the other forecasts made by Sir Christopher was that the use of fuel oil in power stations would, in 1970/71, be less than it would be in 1965/66. In fact, as I have shown, our consumption of imported oil in 1970 was far greater than it was then and it was still rising rapidly.

When he was Chairman of the Atomic Energy Authority, Sir Roger Makins several times emphasised that nuclear power was not a threat to the coal industry. His prophecies were much more restrained than those of his predecessors—and understandably so, since everything that had been said by them on the subject of comparative costs between nuclear power and coal had been proved by events to be wildly optimistic. Makins himself in May 1962 said there was little doubt that the public had been led to expect too much too quickly of nuclear energy. Nevertheless, even his more restrained prophecy that nuclear power stations generating on base load would produce electricity as cheaply as the best coal-fired stations by the early 1970s has remained unfulfilled.

It so happened that a few days after Makins made his forecast I was giving a press conference to cover the Coal Board's report and accounts for 1961. I warned that the argument he had advanced that by the 1970s nuclear power would be competitive with that from conventional stations could prove false. And so it turned out to be.

Neither the Coal Board as a whole nor I personally were against the building of nuclear power stations. Clearly, the country where the atom was first split and which had in many ways originated this new hope for humanity must attempt to develop it and reap the rewards for being early in the field with a power station building programme. Our criticism was that the programme was too big for a new process, and that before proceeding to a large (and, as it turned out, disastrous) programme operating experience should first be gained. The fatal error of too big a programme was made and it stemmed mainly from there being several consortia for power station building. There was never really room for more than one. They were all hungry and had to be fed, and the first programme, based on the Magnox reactors, was much bigger than it need have been for the experimental results that it yielded. Even in the summer of 1971, the building of Magnox stations was not finished. Wylfa in Anglesey was still not in operation.

Sir Stanley Brown, who followed Hinton as Chairman of the Central Electricity Generating Board, was in the early years at some pains to talk publicly about the inter-dependence of coal and electricity generation. Opening in 1962 what was then the biggest power station in the country at High Marnham on the River Trent, he said that the station would take 10,000 tons of coal a day and employ 5,000 miners in producing it. On the same occasion he forecast that coal needed for power generation might reach 100 m. tons by 1980. But for the unnecessarily large and commercially non-competitive nuclear programme, we would now, nine years later, have been a long way on the road towards that figure, in spite of many changes of heart by the electricity industry in the intervening years. There is no doubt that if the

Government and the two industries had worked together in a sensible and sound fuel policy from 1960 onwards, the anxieties of the 1970/71 winter about the price and future availability of fuel oil would have been avoided.

The coal industry would have been run down at a much slower rate with considerable advantage to costs and morale.

Ministers in their weekend speeches have repeatedly boasted that Britain has generated more electricity from nuclear power than any other country in the world. The usual half-truths! What they failed to point out is that not a single unit of all this electricity has been commercially competitive with that from conventional stations built at the same time. So what we have proved with our nuclear power programme is that we can generate very expensive electricity. And that great advance in human knowledge has been obtained at enormous cost in jobs and money, and considerable social disturbance and upheaval.

In the United States as well as in this country, the stations that have been built are all proving more expensive to put up and operate than the forecasts. They have taken longer to build than they were supposed to, with the result that conventional capacity is having to be brought in to make up the gap. Even so, there are serious anxieties in America about the possibility of avoiding further 'black outs' and 'brown outs' because of the failure of nuclear power to fill the space in the power station building programme allotted to it.

Fear of health hazards has caused the abandonment of some nuclear power projects in the States; the German Authorities have insisted on resiting a station; and it was found that another one in Sweden was based on a design that was inherently unsafe.

What went wrong? To begin with, the promises made for nuclear power over the last twenty-five years were wildly over-optimistic. In the early days I heard one learned gentleman in a radio broadcast say that the energy in a railway ticket, if liberated, would be enough to drive trains across the continent of America and back again. Politicians, as well as scientists, became obsessed with the potential of this new power source. There was a kind of Gold Rush atmosphere. In this country, the atom was going not only to solve all energy problems, but, if we were quickly enough off the mark, we would develop a huge export business in nuclear power stations and components.

The wild optimism of some of the scientific correspondents of the newspapers about nuclear power costs had, by the beginning of 1964, begun to look a bit tired. John Maddox, for example, writing in the *Guardian*, while still completely confident that the next generation of nuclear stations would be nearer to competing with conventional stations, had to record that the expected increase in the capital costs of conventional stations had not come about. In fact it had fallen during the previous decade by nearly one half, leading to an unexpected reduction in the cost of electricity based on coal.

By the summer of 1965 it began to look as if the electricity industry itself was having its early optimism about nuclear power stations tempered slightly. Sir Stanley Brown said in explaining why his board had chosen the nuclear Advanced Gas-cooled Reactor (AGR) system for Dungeness: 'There is no question of electricity being cheapened magically by the advent of nuclear power. But the new station will go some way towards reducing our costs. It means that by 1970 we shall have a 10 per cent reduction in the generating costs of 2 per cent of our total capacity.' He added with a surprising touch of realism: 'The golden age has not dawned. We shall not be able to give away electricity.' In fact, Dungeness B is still about three years from coming into full operation and there is no possibility of it coming within a mile of the best coal-fired stations on costs of electricity sent out.

The problem was neatly summed up in February 1967 when Lord Penney, the Chairman of the Atomic Energy Authority, said: 'The country's first nuclear power station programme has not been as successful as expected—partly due to the fact that coal-fired power stations proved more economic, and capital and running costs for nuclear power stations higher, than expected.'

As long ago as October of that year Sir Stanley Brown was also admitting that the first-generation Magnox-type nuclear stations had been uneconomic. (The subsequent corrosion problems suffered by these stations have made them even more uneconomic than he then knew.) But he said that the breakthrough in nuclear power costings had finally arrived. He claimed that his Board knew that it could now generate electricity by nuclear power at a saving of one-tenth of a penny a unit over the cost of the best coal-fired station in the country. In fact, as I have said, there is even now still no nuclear power station producing electricity within 25 per cent as cheaply as that being sent out from the modern coal-fired stations based on the coalfields.

Another of Brown's forecasts at this time also quickly came unstuck. He expressed confidence that every year would bring cuts in the nuclear costings as design improved. Events again have been unkind to him. The latest stations to be started are proving enormously more expensive to build than those that preceded them.

I didn't take too kindly to this 'breakthrough' statement by Stanley Brown, because a Study Group consisting of representatives of the CEGB, the Atomic Energy Authority, the Ministry of Power and my own people was still at work on the subject of nuclear power costs. I was speaking at a meeting in Manchester the day after his claims were reported and I entered the arena with a public rebuke, saying it would have been better to have waited for that Group's conclusions before he made his claims. None of the new AGR nuclear power stations would be operating for at least five years and no one could possibly say what their costs, or the costs of the coal-fired stations, would be at that time.

In fact, it is now widely reported that the CEGB will not build any more AGR stations after those already committed. There is even informed speculation that between now and the time the Fast Breeder Reactor is ready (about which more later), this country will use American reactors. The scepticism I expressed back in 1967 about the costs of the AGR's has been completely justified by events. As so often happened, I was fighting to maintain the confidence of the people in the industry. I was joined by others who were also worried about the effect of the 'breakthrough' claim.

Our campaign had plenty of support. Some probing Parliamentary questions by miners' MPs soon established the facts about the increases in costs that had been experienced with the nuclear power stations then completed and still being built. They not only put down questions to establish the facts of the programme, but also wrote many letters to the newspapers. In June 1967 six of them had a letter published in *The Times* pointing out that the United Kingdom Atomic Energy Authority had been spending in that year about £30 m. on civilian nuclear reactor work, of which about £6 m. was directly attributable to the first two stations in the AGR programme. The royalty payable to the Atomic Energy Authority for this work covered only a fraction of the total cost. They argued, too, that the extra capital required to build nuclear stations, by comparison with capacity based on coal, involved a loss of interest earnings that would have accrued if the money had been put into other activities which would easily earn a higher rate than the 8 per cent that was charged for purposes of calculating the costs of power sent out from the stations. They further stressed that if all these charges were shown, for example, against the costs of Dungeness B, the likely price of a unit of electricity sent out would be at least 0·67d., compared with the official estimate of 0·511d. This estimate made by the MPs is likely to be proved to be much nearer the truth than the official forecasts, but even they may have under-estimated. Construction costs have soared by at least £50 m. The contractors have paid a small part of this bill (International Combustion £7 m. and Fairey £3½ m.) and the CEGB have footed the rest.

As for the station's operating costs, no one now will even attempt an estimate. In fact, there has been an attempt to clamp down on information about this. In the days of the Labour Government, MPs who put down Parliamentary questions about this and other power stations were given answers. Thus, the escalating costs at Dungeness B were revealed and comparisons with contemporary coal-fired stations could be made—which was obviously a great embarrassment to the CEGB.

Since the change of Government, however, MPs have been told in reply to their questions that such matters 'are for the Chairman of the Board who will write to the Member'.

I've seen some of the letters sent to MPs. They give no information at all. In fact, about the only figure they contain is the date.

This conspiracy of silence is a sinister development. Parliament is being denied information of the greatest national importance —information that is essential for seeing whether the vast sums of money being invested by a nationalised industry are being wisely spent.

Back in July 1967, when I made my usual speech to the conference of the National Union of Mineworkers at Eastbourne, I criticised the cost to the country and to the mining industry of building so many nuclear stations instead of mixing in conventional units. I pointed out that when the programme was complete, 14 m. tons of coal a year would be displaced, making a total of 28,000 miners redundant. (In fact, because of the corrosion problems already mentioned affecting the Magnox stations, they are having to be run at reduced capacity, producing rather less electricity and thus displacing less coal.)

Turning to the additional capital cost involved in the nuclear programme, I said the nine Magnox stations would cost £750 m. compared with £225 m. that would have been necessary to construct coal-fired capacity. I went on to show that the £525 m. of extra capital needed to build nuclear stations would have been enough to provide, for example, five more universities *and* 20 hospitals *and* 150 schools *and* 200 extra miles of motorway. This had been the staggering sacrifice that the nation had made to gain a comparatively modest amount of nuclear power technology.

The AGR programme was going to cost about £760 m., or £300 m. more than similar capacity using conventional fuels. Already, that meant that the British people would have to forego another £300 m. worth of houses, schools, hospitals and roads—and this at a time when capital was very scarce for all purposes.

On the assumption that the programme could have been halved without losing any of the knowledge gained, about £400 m. of urgently needed capital could have been devoted to other uses.

I had already been criticised in many quarters for presuming to question the economics of the AGR programme. My justification was the half-million people then on the coal industry's payroll—people whose future would be seriously jeopardised if further mistakes were made in the choice between nuclear power and coal.

A few days before making that speech, I and some of my colleagues had given evidence to the House of Commons Select Committee on Science and Technology, which was considering the nuclear power programme. Leslie Grainger, our Scientific Board Member, with his wide experience in the Atomic Energy Authority before joining the Coal Board, helped to make sure that our evidence, written and oral, was authoritative and unchallengeable. We pointed out that the lost investment opportunity represented by the extra capital ought to be charged against the costs of the programme. We also drew attention to the hidden subsidy represented by the decision not to charge against the cost of the stations the full research and development costs for work done by the Atomic Energy Authority in support of the AGR programme. Furthermore, we pointed out that it would be foolish to assume that no further technical developments would be possible in conventional stations. We ourselves were conducting some highly promising work on fluidised combustion (as I have described in Chapter Four). We completed our written evidence with what I thought was the reasonable suggestion that instead of building all the stations envisaged in the AGR programme, some coal-burning stations should be built to allow for fair comparisons between the two fuels. And we drew attention to a little-known fact. Most people had assumed that nuclear power would not constitute any burden to the balance of payments. We pointed out that about 10 per cent or more of the the costs of a nuclear power station would be in foreign exchange to pay for the imported uranium ore. This was equivalent to a cost in foreign currency of about 10s. for every ton of coal displaced.

Throughout the rest of 1967 and the early part of 1968, Members of Parliament from mining areas kept up their barrage of Parliamentary questions. There was some evidence that Ministers were beginning to get rattled by this pressure. In July 1967, for example, Richard Marsh, then Minister of Power, told Michael McGuire, the Member for Ince, that Dungeness B was still expected to produce cheaper electricity than coal-fired stations. This was either sheer stubbornness or he had been badly misled, because it was obvious even then that the station had no hope of ever being competitive. The Coal Industry National Consultative Council, which represented not only the National Union of Mineworkers and the National Coal Board, but also the management unions and the underofficials' union, requested the Minister to hold an independent inquiry into the technical and financial aspects of the nuclear power programme. This was among the recommendations made by the Select Committee to which we had given evidence that summer. The Select Committee had also criticised in their Report the Minister's delay in sending to them documents prepared by his Department setting out the basis of calculation of nuclear power costs.

When the Select Committee Report later came to be debated

in the Commons, its Chairman, Arthur Palmer (Labour), complained that Marsh had promised in June 1967 to give the
information the Committee were seeking on comparative nuclear
and conventional costs. Palmer disclosed that it was only after
the Committee had brought into play the full powers vested in it
by the House that the papers were finally placed before it. The
fact that a Minister had had to be compelled to give this basic
information attracted astonishingly little attention in the newspapers.

The leading Conservative member of the Committee was the
highly-respected Sir Harry Legge-Bourke. He described me as a
good, robust witness standing up for an industry of which I was
proud. Sir Harry said he accepted that East Midlands coal and
nuclear power were highly competitive with each other.

I am sure that this would also have been the conclusion of the
'outside examination by an independent agency of all financial
aspects of the costing of all methods of electricity supply' called
for by the Committee but never set up.

In November 1967, the Government's White Paper on Fuel
Policy was issued. I deal with this in the next chapter and for the
present it is enough to say that, rather than ending the arguments,
the White Paper simply stimulated them. Barely had the document
been published than sterling had to be devalued and this, of
course, immediately affected the cost comparisons in favour of
coal because it was the only fuel that involved no foreign exchange.

Of course, the enthusiasts for nuclear power were not silent
during all this time. Some absurd claims were made for the AGR.
For example, there were inspired stories in the newspapers that
the Chinese were interested in buying a British reactor of this
type and at one time someone even went so far as to suggest that
the Americans were also in the market. These wildly optimistic
statements have never been justified. No AGR has been sold
abroad and it is quite clear that none ever will be. In this respect
it has been even more of a failure than the Magnox design—two
of which were sold abroad. One went to Japan, the other to
Italy. Warnings about the corrosion problems that affected the
British stations have had to be sent to both countries.

The American-type reactors, because they were smaller,
proved much more acceptable in the export market. Compared

with our overseas sales of the two Magnox-type reactors (both a very long time ago) and no AGRs, the Americans have a great number either in operation, under construction or on order overseas. Twelve countries have bought reactors from the United States, only two from Britain.

So the pay-off for our huge investment in nuclear power stations is a very long time coming from abroad, as well as from the stations we have built ourselves.

The struggle for exports had been a big factor in the decision to build an AGR station at Dungeness. A Government White Paper published in 1964 attempted to reconcile differences between the Atomic Energy Authority and the Central Electricity Generating Board. The AEA favoured its own Advanced Gas-cooled Reactor, while the CEGB wanted to import under licence the American Water Moderated Reactor. The White Paper had something to say for both and faint-heartedly concluded that the Board should shop around.

The White Paper did not please a lot of people, including Lord Coleraine (formerly Richard Law, Conservative MP). Quite rightly he said its effect would be to create more uncertainty, cause more delay and add greatly to the already existing confusion.

Later, the CEGB published their appraisal of both reactors in relation to Dungeness B. A decision in favour of the American reactor would have been a catastrophe for the British nuclear industry with its still-cherished dreams of important export business. Predictably perhaps, the decision was in favour of the AGR. But the appraisal looks very tatty now. It is clear that the comparison was never fairly made. And for reasons I have already given the first station is proving a bad advertisement for the British industry. It is taking about twice as long to build as it should have done, is costing twice as much as was expected, has caused devastating losses to the firms involved in its construction and will never get within miles of the best coal-fired stations on the cost of the electricity it produces.

Even up to the spring of 1971, new defects in the Dungeness B station were still being discovered. For example, the central core of the reactor had to be rebuilt because tests had shown that the material used in some brackets could corrode. All this will add still further to the cost of the station, although it was said at the time that the six months this job would require would not further delay the introduction of the station, which was then already three years behind schedule. Still, many of the American public utilities who have bought the rival reactor have fared almost as badly.

The failure of Dungeness B has meant that the long-awaited day when nuclear power finally becomes competitive has to be postponed yet again.

Ironically enough, just about the only nuclear power station in the country that was completed on time was one in which I had been involved. Before being appointed to the Coal Board I had, for about a year, been Industrial Relations Adviser to Atomic Power Constructions Limited, the main contractors for the Magnox nuclear station at Trawsfynnydd in Merionethshire. I was engaged by Colonel George W. Raby, Chairman and Managing Director of APC. Here I was to meet one of the great unsung men, George Raby's public relations man, W. J. Condren. A good photographer and publicity man *par excellence* and as thorough-going a humanist as ever stepped into shoe leather. How the men respected him!

When I took the job I was condemned by Ted Hill, the Boilermakers' Trade Union leader, and defended by Dick Wood, the Conservative Minister of Power. The job was no sinecure. Many unions were involved, which is always a potential source of difficulty. The site was difficult and isolated. However, we were able to establish proper conciliation procedures and I spent a great deal of time amongst the men, most of whom lived on the site. In the temporary wooden town we created a 'dry' area and a 'wet' area, and quiet areas where men could read or write, and this helped things along a good deal. The company under Raby were prepared to spend money on the facilities that I asked for to make conditions on the site reasonable.

We also produced a site newspaper which, in some ways, was the forerunner of *Coal News*. Of course, the contractors saved a great deal of money by completing construction on time. And the excellent record of industrial relations was quoted by the Minister of Labour as a model.

The only stoppage of work that puzzled me was one which lasted three or four days. It ended suddenly when all the men went back to work without asking for anything and without giving any reason. Subsequently I was having a drink with a few of them and asked what the strike was all about. The answer was quite simple. It had rained without stopping for about a fortnight and the men said they were 'bloody fed up'.

Having learned bitter lessons from the Magnox programme, Governments should have been more realistic when it came to ordering the second series of stations based on the Advanced Gas-cooled Reactor. In fact, precisely the same mistakes have been made again. No AGR station has yet worked commercially anywhere in the world. Yet we have gone ahead and placed orders for five of them at a capital cost of at least £800 m. Mistakes are perhaps inevitable with a new technology. Since the AGR stations are much bigger than the Magnox, mistakes with them are going to be even more costly. All the more reason why we should make only one mistake in the programme, which would have been the case if the first station had been built and operated before the others were commissioned.

The 1967 Fuel White Paper came round to the point of view that we had constantly been stressing. It said: 'No commercial-scale AGR is yet completed and operating, and in a young technology the risk of disappointment must exist.' Alas, the authors of the White Paper were not wise enough to act on their own advice. They went on ordering new AGR stations.

Nuclear power has certainly been given every possible advantage by Governments. As I have said, much of the Atomic Energy Authority's expenditure was on work in support of the civilian nuclear power programme. The CEGB refused to meet all these research and development costs, and after a long and bitter argument Tony Wedgwood Benn, then Minister of Technology, announced that the AGR stations would pay a royalty of 0·014d. per unit sent out, though it was officially acknowledged that this did not represent anything like the total cost of the work done by the AEA. After work had already started at Dungeness B, two more AGR stations were authorised —Hinkley Point B, which in fact is almost certain to start operating first, and Hunterston B in Scotland.

The fourth station in the programme was the subject of a most bitter public quarrel between myself on the one hand, and the Central Electricity Generating Board and the Government on the other—so bitter that it made everything that had gone before look like a tea party.

The proposal was to build a power station at Seaton Carew near Hartlepool, bang on the Durham coalfield. I realised, of course, that if this were a nuclear station the end of coal-fired electricity stations in this country could not be long delayed. Furthermore, it would be a terrific blow to the morale of the industry and to the spirit of the Durham miners, who had always been so loyal to the industry, to the Coal Board, and to the Labour Government which had been elected in 1964.

Durham was one of the oldest coalfields and for more than one hundred years the pits in the west of the county had produced some of the world's finest coking coal, but they were rapidly being exhausted. The only pits with any real future were the coastal pits, which gave access to millions of tons of coal lying in virgin seams under the North Sea.

Not only that, but the Coal Board had invested millions of pounds of public money in developing existing pits and sinking others. This investment was made, not upon what the Coal Board themselves estimated the demand to be in the years ahead, but on actual demand forecasts made by the gas and electricity industries. This was very conveniently ignored by the senior civil servants and forgotten by the various Ministers of Power, particularly those in the Labour administration. If this investment of public money was not to be wasted then a power station at Seaton Carew was very desirable.

Lord Hailsham had become the Minister responsible for the North-East and, properly equipped, as he thought, with a cloth cap, toured the Area. He was pressed very strenuously by the Labour MPs from the North-East and the miners' leaders to agree to a coal-fired station in Durham in view of the serious employment problems facing the county. Confusion then abounded.

In February 1963, the Minister of Power, Richard Wood, announced that no new power station was to be built in Durham. However, miners' leaders shortly afterwards claimed that they had been told by Lord Hailsham during one of his tours of the region that the coal power station was definitely on. The only question was when it would be started.

This confusion was aggravated even further when a CEGB official announced in April of that year that the Board had never intended to build a station there and one would only be planned when the load justified it. To leave such an important announcement, especially when it apparently cut across a promise made by a senior Minister, to a Board official was typical of the electricity industry's frequent insensitivity on matters of social importance. Here was an area hungry for jobs and yet the CEGB were making it clear that their policies were going to be based on only one thing—costs. They argued that to generate the electricity in the coalfield and transmit it to the areas where the demand already existed would be uneconomic. Of course, it is also uneconomic for the country as a whole to be paying men unemployment benefits when they might be at work producing economic wealth for the country.

The CEGB finally announced, after all the procrastination, that they were going to apply for a nuclear power station at Seaton Carew in March 1967. Naturally, they were quite confident that it would be cheaper than a coal-fired station. I doubt whether any outside observer qualified to express an opinion could now be found to agree with them. The shrewdest decision would have been for the CEGB to have opted for a coal/oil fired station on this site. They could then have played the market, using oil when prices were down and coal when oil lost its price advantage, as it did at the end of 1970 and beginning of 1971.

The Durham Miners' Gala always attracts top-ranking Labour politicians and in 1967 it was also the setting for a secret meeting involving the Prime Minister, the Chancellor of the Exchequer (Jim Callaghan), the Minister of Labour (Ray Gunter) and myself. The Union were represented by Sid Ford, Will Paynter, and Alf Hesler, the Durham Miners' leader. The subject of this secret meeting was the power station that was to be built at Seaton Carew. Richard Marsh, the Minister of Power, was not present simply because he was not a speaker at the Miners' Gala and all the others were.

The miners' leaders were incensed that a coalfield like Durham, that had suffered so much from pit closures, should have to face the imposition of a nuclear power station which would close the doors to employment in the Durham pits for their sons still at school.

After all, since the end of 1960 they had seen 50 out of their 109 pits closed and the manpower reduced from 87,000 to 52,000. Even to suggest putting a nuclear power station on the coalfield was regarded by the miners as the kiss of death and an act of treachery on the part of the Labour Government whose support in Parliament had for years included representatives from every mining constituency in the county.

I ought to add, that I was not there at the request of Prime Minister Wilson or any of the other Ministers. Because they were the miners' guests at the oldest of Labour's demonstrations, they were cornered and had no alternative but to discuss the subject of the power station. I was invited by the miners' leaders and I flew up that Saturday morning and with police assistance reached the County Hotel in the heart of the city in good time.

It was a good meeting. Harold Wilson treated the subject with deep interest and the miners with great courtesy. Callaghan and Gunter were obviously concerned about the social and economic problems if Seaton Carew were not to be coal-fired. When the meeting finally broke up the miners' leaders were happy and content. For that matter, so was I.

Yes, the miners' leaders came away fully confident that the Prime Minister and his colleagues had finally recognised the force of their case. But I am afraid this was not the first time, nor was it the last, when Mr Wilson's hearers completely misunderstood his intentions.

Early the next year I made a bid for the station to be coal-fired that thoroughly embarrassed both the Government and the Central Electricity Generating Board. We had been told that, to be competitive with nuclear power, coal would have to be delivered to a power station at $3\frac{1}{4}$d. a therm. We promptly worked out a scheme whereby we could do exactly this. I even offered, as I put it at the time, to throw open the books and let the CEGB crawl all over our figures and examine them with us. This really stirred the pot. The CEGB suggested that we would make them pay more for coal supplied to other stations. We have never had a commercial contract with the CEGB, as I have shown, and they were unwilling to consider one for a particular station. This gave us the chance to reopen the old question of having such a contract, but no progress was made. The nuclear power interests must have been considerably put off their stroke when, having made some impressive-sounding claims for the reliability of the nuclear stations then in operation, there was a dramatic break-down in the AEA's reactor at Chapelcross. It was out of action for many months. Repair work cost £400,000, but the loss in revenue from electricity sales was more than £2·3 million.

Our bid to supply coal at cheap prices on a long-term contract, as I expected, got a pretty cool reception from the electricity industry. Sir Ronald Edwards, the Chairman of the Electricity Council, even went so far as to say that if they didn't have a nuclear station on the site he didn't think they would need a station at all. If we could supply fuel at a little over 3d. a therm they would rather use it in existing power stations to replace some of the more expensive coal they were having to burn there. This meant that the CEGB would be closing our pits, since they had a buying monopoly for electricity generation.

The row over this station flared up again at the end of March 1968 when it was disclosed that the CEGB had already, before getting authority to build a station, committed £1½ m. in an astonishing gamble that the decision would go the way they wanted it in the end. The public relations of the Electricity Board on this subject were amazingly bad. For example, a local official was allowed to say publicly: 'We have no intention of building a coal-powered station.' Could anything have been more likely to infuriate the miners and their representatives at Westminster?

Of course, I just did not believe that the CEGB had committed all this money without some official encouragement. If I am right, and the decision had already been taken, what was going on between the Minister, the NUM and the local mining MPs was never more than shadow-boxing on his part.

The fresh uproar obviously dismayed Marsh. The following day he announced publicly that he was still not satisfied there was any need for a power station at all on the site. He said that he would not give the go-ahead until he was convinced that it was the right time to spend the money and that the demand was there. Confronted by the amazing commitment of the CEGB, the Minister said it was perfectly normal for them to prepare sites in advance. As I say, I believe that in fact the CEGB were not acting completely on their own initiative.

It was beginning to look, despite all the favourable public arguments, as if we were in great danger of losing the decision. So we made another dramatic bid to get the station on to coal.

Our offer of cheap coal had been based on a careful appraisal of the prospects and likely costs of the pits we had picked to supply it.

So I challenged the CEGB and the Ministry to have an independent inquiry into the cost per unit between coal and nuclear power for Seaton Carew. The CEGB had not even asked us to give a quotation, but had gone straight to nuclear power on the basis of their own figures. Of course they could not stop us from putting in a bid. I offered to throw open the books not only to the CEGB but to any independent experts to satisfy themselves about the soundness of our offer. But it was never taken up. Clearly the other side had no confidence in their own claims. A few days later the newspapers were full of stories, patently officially inspired, that the decision was going to be in favour of nuclear.

In fact, the decision was not announced until six months and two Ministers of Power later.

In the meantime, as one Durham miners' leader said, there was 'hell on'.

Bob Woof, the miners' MP for Blaydon, managed to get an adjournment debate in the House. It was a remarkable occasion. In the thirty minutes or so allowed, no fewer than seven MPs, including some Tories, managed to protest about the delay. Put up to reply for the Government on that occasion was Reg Freeson, Parliamentary Secretary to the Ministry of Power, who claimed that the Department were still waiting the outcome of the study report.

One of the speakers in this debate who was cross with the Government was the Labour Member for the Hartlepools, Ted Leadbitter. But his reasons for being angry were quite different from Woof's. He wanted a nuclear station and resented the delay in getting on with it.

Leadbitter had been one of the people on the sidelines during this contest, shouting the odds for a nuclear power station, which didn't make him very popular with the Durham miners. He had taken it upon himself to accuse me of deceiving the miners in Northumberland and Durham. His argument, as far as I could understand it, was that manpower in these coalfields was going to come down. That was true—everyone, and particularly the men, knew it. Leadbitter seemed to think, though, that this was a reason for not opposing a measure that would ensure an even more violent rundown.

He accepted uncritically the CEGB's case for a nuclear station and lost no opportunity of trying to discredit our proposals. The electricity people were very fond (and, indeed, they still keep trying it on) of quoting what a high proportion of their total capacity was coal-based. Leadbitter showed his ignorance of energy matters by making great capital out of this in an article in the local newspaper, the *Northern Echo*. The important thing in deciding how much coal the electricity industry uses is not the proportion of its capacity based on that fuel, but the intensity with which those stations are used. If the continuous base load is given to the nuclear stations (which cannot hope ever to compete otherwise) and oil, while coal is used only for the peak load, the tonnage will be far less than the coal-fired proportion of capacity would suggest. It would be perfectly possible for three-quarters of the power stations to be coal-fired and for them to produce only half the electricity generated.

Leadbitter made the mistake of regarding the forecast that output would be down to 80 m. tons by 1980 as having come from me. In fact, it was his own Government's estimate that I had simply quoted. He rejoiced that one of the aluminium smelters was to be built in the North-East and that nuclear power would be the best way of producing the cheap power it would need. Alcan, with their greater knowledge of these matters, decided otherwise and Leadbitter lost the argument over that, too.

It was suggested that I should argue with Leadbitter on television but I was quite glad to leave him to Alf Hesler, Secretary of the Durham miners, who took him apart in an article in the same newspaper.

Altogether Leadbitter managed to provide some colour in an otherwise grim episode.

Richard Marsh had a miserable time of it in the arguments over the power station. On the one hand the miners' MPs gave him no peace and were pulling him in the direction of coal. On the other end of the rope were the CEGB and, probably, the Ministry of Technology, trying to drag him to nuclear. I appreciate that it was a difficult decision for a young Minister to have to take and I am not questioning his undoubted courage when I say that it must have been with some relief that he hinted in February 1968 in a speech at Newcastle-upon-Tyne that the station might well be postponed. He used the curious argument that it could not burn a pound of coal before 1972 at the earliest. This was a naïve attitude. If he had announced that the station was going to be started and that it would burn coal, we would have had no difficulty in keeping pits in business until the time that this lusty new customer would start consuming their output.

But Marsh was to escape the final decision and it was Roy Mason who on 21 August 1968 announced that he had given the Government's assent to the CEGB's application to build a nuclear power station.

He had been in office not quite seven weeks and he chose to make this statement when the House was in recess and he therefore could not be criticised by the MPs who had been pressing so long for the station to be coal-fired.

I doubt if he had been in the Department long enough to give this important decision the study it justified. It was admitted, for example, that the Coal Board's offer to supply coal at 3¼d. a therm had been ignored and, for the purposes of cost comparison, the average costs of supplying coal to power stations in that area of around 4½d. a therm had been used. It also seemed to us that the capital cost of a coal-fired station had been inflated far beyond the costs that were then being experienced for the construction of coal-fired stations in other parts of the country. This distorted the comparison to our disadvantage.

Mason's decision meant that about 10,000 mining jobs would be lost in Durham, an area where unemployment was already high, and where there had already been a very great number of pit closures. During the summer of 1970 Mason, now in Opposition, had the nerve to criticise the conversion of some coal-fired stations to dual firing. This was being done because it was clear that there would not be enough coal to keep the power stations operating during the coming winter. These decisions did not cost a single miner's job. Naturally, I said in public what I thought of him.

His decision about Seaton Carew was conveyed to me by telephone whilst I was in the coalfields a few hours before the announcement was to be made publicly. I told him of my disappointment but said if that was the firm decision of the Government, then we had lost the fight and I would say no more. His relief seemed evident, even over the phone. But I saw no point in continuing the battle and I had no personal quarrel with Mason.

However, there were much more serious effects than the personal reputation of one Minister. Because the coal industry had been clearly winning all the public arguments about the new power station in Durham, the shock of losing the only thing that mattered, the decision, was all the greater, especially as it had been taken by a Labour Minister and an ex-miner. The effect on the industry's morale was shattering everywhere, but especially in Durham. The miners and the management began to despair of ever getting fair treatment from any Government and, in my opinion, confidence has still not been fully restored three years later.

And even for the advocates of nuclear power, the Seaton Carew station is already looking sick. After construction had started, design changes had to be made for safety purposes on the insistence of the Nuclear Power Inspectorate. Obviously, to change the design after construction had started is bound to be expensive, and this station may well also fail to compete on price with conventional stations.

The Labour Government's action in announcing their decision during August, knowing that it could not possibly be raised in the House until October, produced a characteristic comment from Tom Swain, the outspoken MP from Derbyshire. He said bitterly that he had never expected that 'the Government of our own choosing would treat the miners of this country in this way'.

As for me, I had to try to rally the industry by saying that the decision would only stimulate us to work for higher productivity. 'We are not dead yet,' I said, and added that there was no point in fighting guerrilla actions now the battle was over. But there was no concealing the fact that this was a devastating reverse.

It was galling, a few weeks after the Seaton Carew decision, to see a report by the Brookings Institution of America recommending Britain's investment programme for nuclear power should be substantially reduced. The Brookings Report was written by a team of very highly respected American and Canadian economists. I am convinced that if there had been an independent inquiry, as suggested by the Select Committee and the coal industry, its conclusion would have been the same and it might have been available before the decision on the new station was finally committed to the great harm of the coal industry.

In the House of Commons one of the most enthusiastic supporters of nuclear power was Eric Lubbock, the former Liberal MP, now Lord Avebury. He and I clashed very sharply when I gave evidence to the Select Committee on Science and Technology, of which he was a member.

Lubbock continued the argument through the correspondence columns of the *Financial Times*. Naturally, I pursued him there as well. The main subject of our row was the estimated costs of electricity sent out from the Dungeness B nuclear station when it came into operation. The published figure at that time was 0·51d. per unit, but I knew that a Working Party consisting of representatives of the Atomic Energy Authority, the Electricity Council and the Central Electricity Generating Board had thought it would be prudent to base the calculation on more cautious 'ground rules' assuming, for example, a more conservative interest rate and a more realistic uranium price. This led them to the conclusion that the cost per unit sent out would be 0·56d., even without the royalty which would have to be paid to the AEA. Lubbock, well informed though he usually was on nuclear power matters, was obviously unaware of this new estimate. Events have proved, of course, that even the higher figure adopted by the Working Party has been far outstripped. Lubbock also criticised me for saying that the capital cost of nuclear stations was about double that of conventional stations. He claimed that when Seaton Carew and Heysham, the next stations in the nuclear power programme, came to be built, the gap would narrow very substantially. In fact, Seaton Carew's estimates claim no improvement over those for the preceding station, Hinkley Point B, and Heysham has proved to be half as expensive again as Seaton Carew. So the trend is exactly in the opposite direction to that forecast by Lubbock.

What caused the row between Lubbock and me at the Select Committee hearing was that he refused to accept the higher figure for the costs for the AGR stations that I had quoted. He interrupted me when I repeated it and I had to ask him if he thought I was a prisoner in the dock.

Oddly enough, I had three years earlier in Oxford given the Maurice Lubbock Memorial Lecture, dedicated to the MP's distinguished father.

Later, when another Member of the Committee, Tam Dalyell

(Labour), tried to argue that they had the power to compel my attendance, I sharply pointed out that I was certainly there as a guest since I was a Member of another place (meaning the Lords) and I had come only because I wanted to.

My claim, which had also been made by miners' MPs, that nuclear power was being unfairly helped in the struggle against coal became even more respectable when the House of Commons Public Accounts Committee published their annual report in August 1968—the month the Seaton Carew announcement was made. They concluded quite clearly that the taxpayer was subsidising the development of nuclear power, making a fair comparison with conventional stations impossible. The report declared: 'The generating boards are being helped both with the Magnox stations, on which the nuclear power programme is at present based, and with the Advance Gas-cooled Reactor stations which will be built between 1970 and 1975.' The Committee blamed the Treasury for waiving the generating boards' obligation to pay the Atomic Energy Authority a royalty and argued that, since the last three stations in the Magnox programme were expected to be competitive with the latest coal-fired stations built away from the coalfields, they believed that the royalties should be paid. The Treasury were further criticised for their decision that the royalty on all AGR stations commissioned before 1975 should be paid by the boards at the rate of only 0·014d. per unit of electricity sent out over the full operating lives of the stations. This, the Committee calculated, would yield a total royalty payment of only about £26 m., whereas the final costs of the AEA in developing the AGR system were expected to be about £110 m.

In fact, at least two of the assumptions of the Public Accounts Committee, who acted on the best advice available to them at that time, have proved wrong. For a start, the last three Magnox stations are not competitive with coal-fired stations; secondly, there will be precious little electricity produced at all from AGR stations before 1975.

So now we have the situation where an all-party House of Commons Committee had shown that our complaints of unfair comparisons were justified, adding their weight to the Select Committee on the Nationalised Industries which, some time before, had recommended an independent inquiry. It says something about the way Government is run in this country that the views of these authoritative bodies were blandly ignored. Surely, if we had our values right, this scandalous squandering of gigantic amounts of public resources ought to have caused Governments to collapse. As it is, no one apart from the coal industry was very bothered and even now, three years after the Public Accounts Committee reported, this injustice has still not been put right. Furthermore, although the CEGB publish in their annual reports the price they pay for coal and oil, no information is ever given about the costs of making and re-processing fuels for the nuclear reactors. No outsider has ever seen the terms on which the fuel is bought from the AEA. Yet in America such information is published. The Public Accounts Committee drew attention to this, too. The whole subject of nuclear power costs in this country screams for independent investigation.

Coal was unfairly treated in other ways. Under the relentless pressure we set up for accurate cost comparisons between coal and nuclear power, the electricity industry and the Ministry developed their own neat way out. The earlier comparisons assumed that the nuclear stations would have a working life of only twenty years. When it became apparent that on this basis they could not hope to compete with coal, the rules were simply changed. It was calmly stated that their life would be not twenty years, but twenty-five years. This meant their capital charges could be recovered over a longer term, which would, in turn, favourably influence the cost of the electricity they would send out. What possible justification there could be for the change I still don't know. As I have said, no AGR station has operated on a commercial basis anywhere in the world. But of course the nuclear enthusiasts have the advantage that it will be well over twenty years from now before anyone will be able conclusively to prove them wrong.

The Public Accounts Committee of the House of Commons disclosed a second loss to the taxpayer attributable to nuclear power. Their report issued in August 1969 said that a loss of £6 m. had been the outcome of an agreement between the AEA and the Treasury on the disposal of surplus uranium ore. About twelve years earlier the AEA had bought stocks of uranium ore at prices which were very high because of strong world demand.

When the nuclear power programme slowed down in this country the high-priced stocks proved to be greatly in excess of what was needed. Soon afterwards, the world price of uranium fell and the agreement made available to the generating boards uranium ore, not at the price actually paid for it, but at the world price then ruling. And yet despite all the tricks and distortions, cheap nuclear power is still no nearer.

After the now-notorious delay at Dungeness B, the first AGR station, the electricity authorities and the nuclear enthusiasts got into the habit of saying: 'Just wait until Hinkley Point B comes along. That one is on schedule and it will be the final breakthrough that will once and for all prove that nuclear is cheaper.' Well, it now begins to look as if the pro-nuclear punters will have to find another runner. Hinkley Point B will be at least one year late because, with the station half built, a design fault was discovered which is having to be rectified.

Not surprisingly, there has been a marked lack of enthusiasm for ordering new nuclear stations here and in Western Europe, despite the advantages to the other fuels brought about by the rapid escalation in fuel oil prices. No new nuclear power station came into operation in 1970 in the countries of the European Coal and Steel Community, and none was commissioned in 1971.

In Britain the electricity industry had a tremendous shock when it came to look at the tenders for the AGR station at Heysham in Lancashire ordered in December 1970. The Seaton Carew station, the contract for which was signed in 1968, was expected to cost something like £92 m. In itself, this was a disappointment because it was virtually the same as the previous AGR station at Hinkley Point. So the improvement in technical know-how was not having any effect on the cost of the later stations. But the Heysham project will cost about £130 m., and it is only of the same capacity as the Seaton Carew station. The initial fuel charge will add another £20 m. or so. With a nuclear station, about two-thirds of the costs are for capital, and fuel accounts for only about one-third. With stations using fossil fuels, however, the proportion is reversed. So with money costing as much to borrow as it does at present, the CEGB did not order another nuclear station in 1971 as they were supposed to do.

The future of nuclear power for the generation of electricity is said to lie in the Fast Breeder Reactor. But here, too, the latest developments are not promising. The FBR is eagerly being sought by the Americans and the Russians as well as ourselves because it 'breeds' fuel, giving the possibility of making unnecessary further imports of uranium. So the potential attractions are immense. However, the design of the FBR is much more intricate than that of the AGR, which was itself more complex than the Magnox.

The technique of this new type of reactor involves working beyond the present state of knowledge of the behaviour of materials. Each series of reactors has produced its unforeseen problems. The Fast Breeders will be no exception. The first British prototype at Dounreay is now a year behind schedule. Already it looks as if the 'doubling period'—that is, the time needed to breed the same amount of fuel as that used up—is going to be much longer than was hoped. If so, all of the hoped-for cost advantage will be lost. The American Fast Breeder prototype suffered a core melt-down—in other words, an explosion.

I hope the problems will be overcome, but it is quite certain they won't be solved quickly. It is quite likely, in fact, that the FBRs will have some importance as a source of fuel, but that they will never produce cheap power. A similar conclusion has been publicly expressed in a recent speech* by no less an authority than Lord Hinton, former Chairman of the CEGB and former Managing Director of the UKAEA Industrial Group.

Without the early experiments in Magnox and now the Advanced Gas-cooled Reactor, it might well have been impossible to move on to the Fast Breeder Reactor, but there can be no denying that both programmes were too large for the scientific function they had to perform. Before the last Magnox station was completed they were obsolete. Before the last AGR is commissioned it is more than likely that it too will be obsolete. What we have seen is another classic example of over-optimism, coupled with determination to go ahead regardless of the consequence to the men who work in coal. Ministers must accept responsibility for authorising the large capital expenditure that made these decisions possible.

The truth is that they were in no position to make the judgments

* Lecture to the Council of Engineering Institutions, 30 September 1971.

and neither were their advisers. Only a permanent Energy Commission on lines I shall describe in the next chapter, looking objectively at all the energy industries and with no other axe to grind, could provide an objective view and impartial advice to Ministers. But this is crying for the moon, as no one is going to part with the authority he already possesses. As a result the country suffers.

For my part I can only regret the slum clearances, schools, hospitals, roads and universities that could have been paid for with the money that has been poured down the nuclear drain.

10

Fuel Policy Fury

Harold Wilson, writing about the work of the Labour Government which he headed from 1964 to 1970, said that, while every Prime Minister's style of Government must be different, he found it hard to resist the view that 'a modern head of Government must be the Managing Director, as well as chairman of his team, and this means that he must be completely au fait, not only with developments in the work of all main departments, including the particular responsibility of No. 10, but also with every short-run occurrence of political importance'.*

On his own rating, therefore, his Government's 1967 Fuel Policy was so bad and inaccurate that he earned his dismissal from the post of Managing Director by the shareholders. Privately owned businesses making such a colossal blunder would have been bankrupted. Many Managing Directors have been sacked for much less. What made things much worse was that Wilson knew a good deal about the industry. His first book was *New Deal for Coal* and as 'chairman' he could not have been unaware that the miners and the mining communities gave him his majority in 1964 as they had given Clem Attlee his in 1945. That Government came to an untimely end following the resignation of two members of the 'Board'—one of whom was Harold himself.

In 1960, as a member of the National Executive of the Labour Party, he was part-author of a statement which said: 'These industries (coal, oil, atomic energy, gas and electricity) must make long-term plans for their development and there must be, as the official Ridley Committee recommended in 1952, adequate co-ordination between them. If not, economic dislocation, much waste of capital and considerable social hardship are virtually impossible to avoid.'

One would have thought that our aspiring Managing Director giving support to such a philosophy in 1960 would, by 1964, when he eventually became the Managing Director, have so

* Harold Wilson, *The Labour Government 1964–1970: A Personal Record*, London: Weidenfeld and Michael Joseph, 1971.

directed affairs through his management team that the evils so well understood and described in 1960 would have been circumvented. Not a bit of it. His failure to ensure that the right people were appointed to see that the 1960 objectives were reached meant that the fuel policy for which he accepted full responsibility, as I did for the Aberfan disaster, led precisely to those things which he was pledged to avoid. There certainly was economic dislocation, from which the country has not recovered; there has been, and still is, much waste of capital (the last of the obsolescent Magnox nuclear power stations built at a total cost of £750 m. is not yet fully commissioned); and there was indeed so much social hardship that men and women who were loyal members of the Labour Party withdrew their support.

But to continue with the prospectus issued in 1960 with the backing of our aspiring Managing Director:

There are powerful reasons for concern at the recent rapid rundown of the coal industry. In the first place, loss of men from the coal industry has been so great that there is already a shortage of miners in some areas. The mere threat of over-production is enough to make many men, who remember pre-war days, leave the pits. If this manpower loss is not checked in a comparatively short time, the nation may once again be short of coal. Moreover, the loss of active younger workers, especially from the coal face, in the long term inevitably threatens the industry with rising costs.

In the second place, an unplanned closure of pits will in some districts simply mean unemployment and hardship. This is because many miners are located in districts where there is no other source of employment.

In the third place, a mine, unlike a factory, cannot be closed because of temporary over-production, and then when demand increases, quickly be re-opened. Often a closed pit cannot be re-opened at all.

It would be short-sighted to destroy coal reserves just because the demand for coal has dropped temporarily, especially when it is remembered that coal is not being displaced because its price is uncompetitive.

Coal is not only a cheap and efficient fuel but it is the one source of fuel supply which the nation has completely under its own control. To rely complacently on imported oil as the Government is doing is to expose the economy to two dangers: foreign supplies of oil might be seriously reduced as they were after the Suez crisis; and the nation, which has struggled for 15 years against repeated balance of payments crises, might be unable to afford ever-increasing quantities of imported fuel.

The statement went on: 'The Government must immediately set a realistic minimum figure for coal production for some years ahead. This is the industry's major need. For, without a specific target, the National Coal Board will not be able to plan sensibly its investment and production programmes, nor will the crisis of confidence that those working in the industry are now experiencing be overcome.'

The statement was issued in September 1960 under the title: 'Fuel and Power, an immediate policy; joint statement by the Labour Party and the Trade Union Congress'. On 1 October 1960 I joined the National Coal Board, the joint statement neatly tucked into my inside pocket.

The men appointed by the Managing Director to head the Ministry of Power had probably never read it, or if they had, they had forgotten it. The Managing Director himself had not forgotten it. As is well known, he has an extraordinarily good memory and forgets nothing.

What went wrong subsequently was that the Managing Director changed the policy, as of course he was entitled to do, but did not consult the plant manager, which by that time was me. In business an unforgivable managerial lapse.

In November 1967, the Labour Government's Energy Policy was issued with a good deal of publicity. The Managing Director had every intention of putting the policy White Paper to debate in the place where many of his finest hours had been spent and where many an economic battle has been won and good industrial relations policies have seen the light of day. This is Parliament, where the Government of the day has an automatic majority, thus ensuring that the Managing Director's policy is carried on all occasions. The reception of the White Paper gave some pause for thought.

The fury of the miners' MPs when it was published was

expressed by Tom Swain, who was chairman of the miners' group in the House of Commons that year. There had been suggestions that Dick Marsh, the Minister responsible, should meet the miners' MPs, but Tom, in his typically blunt way, said: 'If the Minister is coming here to placate the miners' group he has the biggest bloody job he has ever had in his young life.' The members of his group demanded a two-day debate in the House on the subject. A date was fixed, but just over a week after the document was published the Prime Minister yielded to pressure from the miners' MPs and, backed up by some of his own senior Ministers, withdrew a Government motion approving the White Paper. The debate was off and stayed off. The Commons never did debate the policy statement.

The Managing Director evidently felt that discretion was the better part of valour. The miners' MPs did nothing further. A Parliament which found the time to debate and decide whether murderers should be hanged or not, discussed how to be kind to criminals, made abortion easy and permitted pornography and other healthy exercises for the permissive society, could not find the time to discuss the bedrock of its national prosperity. Courage or cowardice—take your pick.

What the miners' MPs did not realise was that getting the debate called off did not mean that the policy was abandoned. Those who wrote it and got Ministers to agree used the policy as the basis for all decisions at official level. The miners' MPs had missed a trick. The Managing Director had cleverly dished them.

The politicians had been defeated by the civil servants, who continued to use the White Paper as the basis for planning, even up to comparatively recent times when events had long since completely disproved it. One man who has clung to the White Paper—probably the only living soul who continued to believe in it—was Dick Marsh. Even the day after Prime Minister Wilson had pulled the White Paper carpet from under him, Marsh was defending it to the Parliamentary Labour Party. The Managing Director had spoken about reviewing the White Paper (and there was good reason to do this, because sterling had just been devalued, altering some of the price comparisons in favour of indigenous coal), but Marsh insisted that the underlying policy would remain the same and there was little that could be done to change the pattern of fuel use. One had to admire his deter-

mination, but not his wisdom. One of the Minister's few supporters in the House of Commons was Harold Neal, then the disgruntled member for the mining constituency of Bolsover in
Derbyshire. Neal had succeeded me in 1951 as Parliamentary
Secretary at the Ministry of Fuel and Power when I became
Minister of Labour. Speaking later in a Commons debate on the
Coal Industry Bill, Neal said that the Prime Minister ought to
protect his Ministers from people like me and suggested that the
only solution would be my immediate dismissal. What had
apparently incensed Neal was a meeting (which I shall describe
later) with the planning authorities and union leaders at which
I had disclosed the effect on employment of the Government's
policy. But it was the policy of his Government colleagues that
was the problem and it was that that he should have been attacking, not my explanation of its consequences in human terms.

During the strenuous in-fighting that went on during the
months preceding publication of the Labour Government's
Fuel Policy in November 1967, it obviously looked to outsiders
as if coal was being battered by the challenges of the newcomers,
nuclear power and natural gas. The *Financial Times* in a leading
article said of me: 'He must feel rather like the unfortunate
Sisyphus, condemned to pushing a stone to the top of a hill only
to see it slide down to the bottom just as he was nearing the peak:
as soon as it is in sight of solving one problem the coal industry
finds that two more have taken its place.'

I don't agree with the final statement in that quotation, but
there was certainly never any shortage of problems and challenges
during my ten years at the Coal Board. But standing up for coal
and arguing the industry's case was never a burden for me. I
knew I had a united industry behind me. I was determined to
do the best I could for the people who depended upon it.

One of the many people who much earlier correctly foresaw
a time when oil would become scarce and more expensive was
General Sir Campbell Hardy, who was appointed Director of the
Coal Utilisation Council in October 1960. On the very day that
I did my first stint at Hobart House he made a speech to the
Coal Industry Society in London saying that he could see a time
when the cold war would become more intense, Russia would be
short of oil, and she would start making her influence felt in the
Middle East. Great care would be needed to safeguard Britain's

supplies. He argued that there should be a national plan to make greater use of our own fuel resources.

Another early and more detailed forecast was made by the Coal Board's Economic Adviser, Dr E. F. Schumacher. In a lecture delivered in Duisberg as long ago as April 1963, he said that all the signs pointed to an energy shortage within the next ten to twenty years. In fact, of course, this has happened even more quickly than he predicted. This seemed to be a remarkable forecast in the context of its time, because oil was then both abundant and cheap in Europe.

Sad to say, the price appeal hypnotised all the Western European Governments and they are all alike in regretting their past policies now. As Schumacher pointed out, shortages of power always have to be dealt with, at whatever cost. There was, then, no justification for abandoning pits and prejudicing future energy resources on the ground that there were transitory opportunities to obtain supplies cheaply.

We joined the other Western European coal-producing organisations in putting on a stand at the International Energy Exhibition held in Paris in May 1964. The whole purpose of our joint display was to show the mining industry modernising and re-organising itself to meet the competition of other fuels.

When I visited the Exhibition for the opening at a press conference attended by about 150 European journalists, I warned that unless the Governments of Western Europe gave enough protection to their coal industries over the next few years, coal would not be available when it was needed. By the end of the century the world's demand for energy would be four times as large as it was then. When the glut of oil and natural gas then being experienced had been swallowed up, a world shortage of energy was probable. For Western Europe to depend for a large part of its energy over the next forty years on foreign countries whose political stability was doubtful would be to court disaster.

If that warning had been heeded and a bigger coal industry had been maintained in Western Europe, it would at least have provided a salutary check on the recent swift rise in oil prices. Dependence on coal imports would have been lessened, and some of the recent and current anxieties over supplies would have been avoided.

Computer-based models of likely energy supply, demand and cost, such as were used in Britain to produce the now-discredited 1967 Fuel Policy, tend to overlook human forces. I and many others could see that it would only be a matter of time before the oil-producing countries would co-operate together to force the consuming countries to pay more for their supplies. The computer, or those who were feeding it, evidently could not.

Whenever and wherever the opportunity occurred, I warned about possible energy shortages and their catastrophic effect. For example, when I addressed the United States National Coal Association Convention in Washington in June 1965, I predicted that world energy requirements might double between then and 1980 and then double again by the year 2000. This would mean, I said, a rise in world requirements from 4,371 m. tons of coal equivalent in 1960 to 9,500 m. tons in 1980 and to more than 17,000 m. tons in the year 2000. I showed that proved world oil and natural gas resources were very small compared with total world consumption in the years to come. The world's reserves of coal far exceeded the reserves of oil.

An added complication of course is that the oil reserves are mainly in the 'have not' countries. The demands of the developing areas of the world are going to grow very rapidly. A few years ago everybody got terribly elated about the discovery of new oil reserves in the Sahara. More recently there was great excitement over the Alaskan discoveries. But when one remembers that to meet just the growth in the American market alone it would be necessary to make discoveries of Alaskan proportions every two to three years, one has to sober up pretty smartly.

In Paris, Washington and London I sounded my warnings. But it is only very recently, after the combined action of the Middle East oil-producing countries, that people have really started to think seriously about these matters.

Our folly in Britain in running down our coal industry too far, too fast, was more than matched in the countries of the European Coal and Steel Community. As long ago as the summer of 1966, estimates produced jointly by the High Authority of the European Coal and Steel Community, and the Common Market with the Euratom Commission, forecast that the demand for energy in the Community would nearly double by 1980. One would have thought, given this assumption, that every effort would have been made to see that as much as possible of this huge energy market should be supplied by coal, the only completely indigenous fuel in Western Europe. Not a bit of it. The forecast made in the summer of 1966 of a doubling in total of energy demands also put forward a target of 190 million tons of coal production for 1970, which represented a cut-back of no less than 30 million tons compared with the 1965 figure.

I remember commenting on this curious phenomenon to a colleague in the French Coal Board, the Charbonnages de France. With characteristic Gallic resignation, my colleague said: 'Yes, what we all suffer from is an up-rush of enthusiasm for euthanasia.' Certainly nothing short of a built-in death wish could adequately explain the enormity of the folly perpetrated upon the coal industries of Western Europe.

This foolishness was not repeated in the United States or in the Soviet Union. Although these countries have diametrically opposed economic systems, they have both reached the conclusion that they must develop their indigenous coal industries to produce far more fuel than they were then mining. And yet both these two giants had reserves of other fuels like oil and natural gas, whereas Europe until recently had none.

Hints about the possible shape of the fuel policy being prepared in the Ministry of Power began to emerge early in 1967. No doubt these were intended to test reactions. Since one of the earliest figures that leaked out was 100 m. tons a year for the future size of the coal industry, reactions certainly came. I for one took my cue—vigorously and loudly.

The air was full of optimism about North Sea gas. Richard Marsh was talking in terms that gave him the reputation of wanting to be the man who brought this new, indigenous source of energy into use. And he chose the odd occasion of a luncheon to mark the presentation of safety trophies to the miners of South Derbyshire to express some of his hopes. The function was held near Swadlincote which, as most people have repeatedly heard, was in the constituency of George Brown, the then Foreign Secretary. Having sat through the pre-lunch speech in which his Cabinet colleague talked about the North Sea producing by 1978 twice as much gas as the entire supply then being consumed in Britain, and having heard his equally enthusiastic description of the plans for nuclear power stations, George

Brown no doubt felt that he ought to try to reassure those of his mining constituents who were present. When he followed Marsh, George said that the industry's position was still tremendously relevant to the solution of the balance of payments problem.

Not to be outdone I got to my feet and drew attention to the rapid increase in the industry's productivity in the preceding six years. And I added, especially for the benefit of our distinguished guests, that if all British industry had done so well, there would have been no problem about the balance of payments.

In the spring of 1967, when Richard Marsh was preparing for publication of the Labour Government's Fuel Policy White Paper, he arranged a weekend conference at the Selsdon Park Hotel in Surrey, which was used three years later by Edward Heath when he wanted a convenient, discreet centre for pre-election discussions with the other Tory leaders.

The venue in our case was supposed to be secret, but of course the Press soon found out about it. In fact the hotel was almost within sight of the home of the then Industrial Correspondent of the *Daily Mail*, Keith McDowall. Our hideout was blown, as they say in gangster films, and the Press photographers turned up to get pictures of the Minister and the Chairmen of all the nationalised fuel industries. We all got mixed up with a wedding party at the lunch break, and wedding guests had a good time showering us with confetti. All the trappings of security were abandoned and we joined in with good humour.

The Selsdon Park documents are still secret, of course, but I can say that the record of our conversations would certainly show that recent trends in the energy market have completely justified the line I took on behalf of coal.

We met on Thursday, 18 May and completed our labours on Saturday the 20th. I had with me Derek Ezra, then Deputy Chairman of the Board, Arthur John, the Member of the Board for Finance, and Leslie Grainger, our Scientific Member, who was a tower of strength in the discussion of nuclear power. He was the only man present with a deep and practical knowledge of nuclear energy, having worked from 1951 to 1962 on metallurgical and other problems in the Atomic Energy Authority, latterly as Assistant Director of Harwell. As a matter of interest—and this supports my continued criticism of the quick changes of the civil service personnel—only two of the high-ranking civil servants, as I write four years later, are still involved in fuel matters in the Department: Robert B. Marshall, now Sir Robert, Secretary (Industry), and Mrs J. M. Spencer, Under-Secretary, Gas Division.

Writing from memory and my rough notes I recollect our argument about the obligation to provide the quantities of smoke-less fuel for the public in 1970 which rested not just on the Coal Board but on the gas industry as well. There was complete acknowledgment by Sir Henry Jones, Chairman of the Gas Council, that various undertakings had been given by Area Gas Boards to local authorities about supplies. He didn't think, however, that there would be much demand for gas coke in the early 1970s. In the event, at the approach of the 1969 winter there was an incipient crisis, and only the mild weather and a fantastic number of conversions from coke boilers to other fuels prevented a real crisis in the winter of 1970. Furthermore, many new Smoke Control Orders had to be delayed and existing Orders suspended.

The argument about the comparative costs of nuclear power and coal-fired stations was fast and furious. In May 1967 we heard by accident that a report had been produced in the preceding January by a Ministry of Power Working Party into Nuclear Costs, consisting of representatives of the Electricity Council, the Central Electricity Generating Board and the Atomic Energy Authority. This committee, which had obviously felt itself qualified to make all sorts of calculations about coal costs without bothering to talk to the Coal Board, had been set up a year before we even knew of its existence. I was naturally outraged when I found that we were not being consulted about a matter of such great importance to our future. Quite obviously we were deliber-ately excluded.

After I had kicked up merry hell we were allowed to join the Working Party and our representative was the best man we could possibly have had—Leslie Grainger, whose great experience of nuclear power I have already described.

We made our criticisms of the Working Party's Report, and after the Selsdon Park conference the new Group, with Grainger on it, considered those criticisms and issued a new Report. Richard Marsh erroneously told the Select Committee on Science and Technology that Grainger, on behalf of the Coal Board, had said he was now satisfied with the cost estimates produced by the Group. To give him credit, Marsh immediately apologised to Grainger when he realised he had been misled.

In fact, we issued an Addendum to the Report setting out our dissent. We pointed out that the cost difference between nuclear power and coal-fired electricity for new stations commissioned in 1970–75 was certainly less than the errors in forecasting (events have more than borne out that statement) and we drew attention to the excess capital that would be required for building nuclear stations.

This Addendum was published in the Select Committee's Report, as was our original Memorandum submitted to the Committee in which we argued, among other things, the folly of making assumptions about the costs of the Advanced Gas-cooled Reactors until one had been operating for several years. Those documents of ours are on the record for everyone to read. They have stood the test of time and can still be read by their authors without blushing—unlike the Reports of the Working Party itself, and most other studies on the subject.

However, the Chairman of the Working Party, a civil servant, was in 1969 appointed Chairman of an Area Electricity Board. He was another of the Ministry officials concerned in fuel policy planning who had departed to pastures new by the winter of 1970/71 when the policy finally collapsed in ruins.

My problem however, apart from the long-term proposals, was the immediate one of the pit closures. The manpower was being run down at the rate of about 25,000 men a year; this was manageable at this stage because we had a virtual ban on recruitment and the natural wastage was extremely high, as might be expected with such a large labour force. As the industry contracted it would become increasingly difficult to maintain that rate of contraction by the same balancing of wastage and recruitment. Moreover, the sort of figures for coal production proposed for the coming years would mean a rundown of 30,000 a year, and on a smaller manpower this would be an incredibly difficult, if not impossible, task.

Labour Ministers have at various times claimed the credit for the generous way redundancy was financed, but it was at the Selsdon Park conference that I advocated paying redundant miners 90 per cent of their wages for three years. I offered to work out the costs for a rundown of between 25,000 men and 35,000 men a year. Any higher rate would be completely unmanageable. What most people never took into account was that our overhead costs per ton rose rapidly with contraction and a further write-off of capital was inevitable.

It was a very difficult period. The Coal Board were obviously going to be the residuary legatees after oil, natural gas and nuclear power. Only the Coal Board had the human and social problems to deal with, and whilst the other industries were plainly sympathetic, they were all growth industries and sitting pretty.

I came away from Selsdon Park tired, dispirited and despondent. By Monday morning, nine o'clock at Hobart House, I had recovered my good spirits and got stuck into the job.

A very nicely worded public statement was issued, which included the following: 'The main purpose of the studies was to establish the cheapest possible pattern of future energy supplies, compatible with the economic, social and technical realities of the situation. In the course of the meeting each Chairman gave an account of the prospects for his industry and there was a full and frank exchange of views.'

Very, very peaceful. You could almost visualise us drifting quietly to sleep. Such are the generalities and euphoria of public statements.

By the end of June newspapers were carrying stories that we had achieved a great triumph in the Selsdon Park discussions and that the intention was now to plan on the assumption that in 1970 the National Coal Board would be able to sell 150 m. tons a year. This, again according to the newspapers, was 15 m. tons above the level suggested by the Ministry when the discussion started.

When the first study documents were produced as a basis for the White Paper, they included a forecast for 1980. This showed coal coming down from the planned 120 m. tons in 1975 to a mere 80 m. tons five years later. As before, oil, natural gas and nuclear were all expected to grow.

These figures were reported in the old-style *Sun* newspaper by Geoffrey Goodman, the Industrial Editor, in July 1967. Four months later when the White Paper was finally published, it was seen to contain no estimates for 1980. This was one concession that I had achieved in all the strenuous battling that went on at the time of Selsdon Park and just after.

During the months between the Durham Miners' Gala in July 1967 and publication of the Fuel Policy White Paper in mid-November that year, the gap between the miners and their leaders on the one hand and the Government on the other was widening rapidly. The Prime Minister was anxious about this of course and he was hoping that the Regional Development and Economic Planning Councils would help by doing two things—finding new jobs for the miners who were going to be displaced, and recommending the deferment of more pit closures to ease the impact on their Regions, thus justifying the Government in paying the costs to the Board of these postponements. It had always been the Board's policy to keep close to the Regional Economic Planning Councils as part of our endeavour to try to dovetail pit closures with the advent of new industry, so as to minimise redundancy and unemployment.

I therefore arranged a meeting with these Chairmen at Hobart House five days before the White Paper was to be published. I also decided to invite Sid Ford and Will Paynter, President and Secretary of the National Union of Mineworkers, because I expected the information I was going to give would get out quickly and I wanted them to have the facts at first hand.

As the dire prospects for each Region were revealed (and we had prepared a huge chart to set them out clearly), I could see how appalled my audience were. I showed how, at the time I was speaking, there were about 380,000 men on colliery books, but this would come down to 270,800 by 1971, to 159,000 by 1975 and, extrapolating the trend if production was to be brought down to 80 m. tons as Dick Marsh let slip out in the House of Commons, to only 65,000 by 1980. (I omitted the figures for the small Kent coalfield, so the totals differ slightly from those in the White Paper.)

Harold Wilson has this to say about the incident in his book:* 'Concern about unemployment had been heightened by sensational and inaccurate disclosures the previous Friday—attributed to the Coal Board—that the coal industry was to be run down from 390,000 to 65,000 men by 1980. I had been forced to run

* Harold Wilson, *The Labour Government 1964–1970: A Personal Record*, p. 453.

the gauntlet of anxiety while touring the Yorkshire coalfield the day they were published.'

The figures were based on output forecasts of his own Ministry and it was Harold Wilson himself who had asked that the Chairmen of Economic Councils should be advised of the situation, as he described twenty-three pages earlier in his own book.

The information remained confidential for at least two minutes after the end of the meeting. A corps of pressmen had gathered outside the room, and as they left some of our guests, vastly disturbed by the figures they had been given, blew the gaff.

The effect was sensational. Ministers whose cherished White Paper formed the basis for the manpower statistics ran for cover. I was attacked by Roy Mason, then Minister for Defence Equipment, and by Ray Gunter, Minister of Labour; Leo Abse, another Labour MP, demanded my resignation.

This was the usual ploy of angry Labour MPs. Their idea of democracy was when we all dutifully sang out of the same hymn book as the one that they were using at that particular point in time. As employers, slightly removed of course, they had only one solution to deal with any employee who offended—to sack him. I suppose they must have inherited this from employing the system of Victorian capitalists, whose way of making money was evidently to be deplored, but whose methods of dealing with recalcitrant employees they admired. Actually I've earned a place in the *Guinness Book of Records* for the number of times my name has appeared on the Order Paper of the House of Commons, when I was no longer a Member. They seemed to forget, if they ever knew, that after fifteen years as a Member of the House of Commons I was quite capable of assessing the importance of them and their actions very accurately.

NUM lodges (branches) in Durham, Lancashire, and Nottinghamshire voted to stop paying the political levy to the Labour Party. There was even talk of forming a separate political party. Conservative backbenchers put down a motion commending me for my honesty and condemning the Government for their failure to provide alternative jobs in the mining areas. Miners near Burnley called for a national strike to fight the Government's pit closure policy, and in Kent the NUM told their MP, David Ennals (a junior Minister), that he should be prepared to resign if necessary to support them. The public brawling caused by my announcement ensured that, when the White Paper was published, it emerged in an atmosphere crackling with tension and drama.

In fact, the manpower rundown would have been much worse but for the arguments I had put forward at the Selsdon Park Conference. When the White Paper was published, it said that the decision had been taken to try to hold the demand for coal at around 155 m. tons in 1970. I don't suppose I shall get any medals for the fact that this was proved to be an extremely sensible decision. We needed every ton of that coal.

After what happened to oil prices in 1970 (an increase of over 60 per cent in twelve months), one can only wince on re-reading bits of the White Paper. This paragraph, for example: 'It is difficult to predict the course of oil prices. There are a number of reasons for expecting them not to increase. The industry is continually searching for ways of cutting costs, as for instance by the use of very large crude oil tankers to reduce freight charges and increase flexibility and security of supply. Competition is strong both between companies and between sources of supply, and the surplus of crude oil seems likely to persist for many years despite the expansion in world demand. Here and elsewhere oil will be up against increasingly strong competition from natural gas and nuclear power. On the evidence available, it seems likely that oil will remain competitive with coal, and that pressure to force up crude oil prices will be held in check by the danger of loss of markets.'

Not only has that been falsified by events, but I know no one today in the oil business who expects oil ever to be cheap again. Indeed many are doubtful about the quantities available against the demand over the years ahead to the end of the century.

The shaking that the oil companies went through at the hands of the producing countries early in 1971 yielded some strange results. For example, Sir David Barran, Chairman of the Shell Transport and Trading Company, said in a speech that the more expensive and precarious oil becomes, the greater would be the incentive for consuming countries to develop alternative energy sources of their own. I know that to some extent he was trying to warn the producing countries not to push too hard, but it was nevertheless a remarkable thing for such an important leader of the oil industry to say. In the same speech to the Fuel Luncheon

Club Barran said something else which emphasises the enormous growth in consumption of oil products and the urgent need for the industry to find more reserves. He forecast that in the 1970s alone the world would consume considerably more oil than it had since the first well was sunk in Pennsylvania 110 years before. The industry would need at least £200,000 m. in capital in the decade and this money could not be found unless substantially higher revenues were earned. No wonder Barran concluded that the buyer's market for oil was over and the moment of truth had arrived for the consumer.

The authors of the White Paper in 1967 also wrote this piece of elegant prose: 'The repercussions of the Middle East war could have significant implications for certain aspects of fuel policy. . . . The Government are conducting a thorough re-examination of ways of strengthening the security of our supplies. They nevertheless consider it right to base fuel policy on the expectation that regular supplies of oil at competitive prices will continue to be available and they believe that it would be wrong to deny to British industry the advantages that oil can bring.' But what could one expect from people who were too busy to find out or too ready to take advice from everywhere except Coal Board quarters?

A very senior civil servant in the Ministry of Power, one of the co-authors of the White Paper, dismissed my references to the newly formed OPEC (the Organisation of Petroleum Exporting Countries) by saying: 'You can forget about them. They will never amount to a row of beans.'

It was much the same phrase that employers used about trade unions in the last century.

He was just repeating what some of the oil men at that time were saying. But I knew the politics of the Middle East very well. I had been there, and in the early months of 1957 had sat in the lounge of the modest home of President Nasser and listened to his hopes and ambitions for the Middle East, which included some Arab States who had oil reserves and some who had not. I didn't take OPEC so lightly and said so, but to no avail.

One of our Welsh miners, with the blue scars of years in the pits across his forehead, hit the nail neatly on the head, when he said in a film we made: 'Do you think the Arabs are going forever to live in tents?'

You don't have to be a university graduate to see the obvious, but sometimes those who know so much, see so very little. It was the Welsh miner who proved to be right, and the highly educated and expensively briefed civil servant who proved to be disastrously wrong. For it is OPEC that has been mainly responsible for pushing oil prices almost out of sight. The Labour Government's assumption that they could plan a fuel policy based on abundant supplies of cheap oil turned out to be calamitously mistaken.

After the Arab–Israeli war in the summer of 1967, which prevented the flow of oil through the Suez Canal, the oil companies set out to diversify their sources of supply. First they turned to Nigeria. But hardly had they started the expansion of production in that unfortunate country than war broke out there.

Next, they turned to Libya. By the time the negotiations with OPEC took place the United Kingdom was getting about a quarter of all its supplies from that source. But the price negotiations with Libya proved to be even more expensive and difficult than with the other producers. The fact is that no amount of diversification can provide any protection against a concerted move by the oil-producing countries of the kind which has given us such a towsing.

Coal has no value of course until it has been produced and brought to the surface. Coal that is stocked on the surface is obviously worth more than coal in the seam where it has been for the past 250 million years. This blinding glimpse of the obvious was not, apparently, something that occurred to the authors of the White Paper. For they said the Government were most anxious that excessive stocking should be avoided. In fact they insisted upon a limit of 30 m. tons. They even worried that the additional stocks might remain unsold indefinitely. It wasn't many years before the Government were wishing they had some of those stocks back again.

In fact, a wise stocking policy would have regarded them as a means of ironing out the booms and the slumps in demand. Jim Bowman, when he was Chairman of the Coal Board, allowed the stocks to rise so that he wouldn't have to close pits so fast that the people of the industry would suffer. He used to say that the 'Mountains of Mourne in the pit yards' didn't bother him. And he was quite right. But, then, Jim wasn't a Minister or a civil servant. Just someone who'd spent his whole life in the industry.

The day will come when a wise Government will do with coal what it has always done with oil, and that is to ensure that there is always a strategic stock held on Government account. That has been the case with oil for many years but that is because the defence of the realm depends upon it. The Service departments had a better idea of the cost of not having fuel when you want it than the civil departments.

I continued to kick up a row about the White Paper more than a year after it was published. Giving evidence to the Commons Select Committee on Nationalised Industries in April 1969, I emphasised the seriousness of the manpower implications of the Government's policy. I showed how the estimate of 120 m. tons of coal production in 1975 would involve an even steeper run-down in manpower than had been experienced in the past. I said the industry was very near the edge and warned that if the slide in manpower turned into a sudden drop, there would be a very serious effect on morale in the industry and on production itself. I said, too, that there was a real danger that we would never be able to recruit the people we needed and, furthermore, there might well even be a refusal to co-operate in productivity drives.

'It might also begin to destroy the very good industrial relations built up—and it may be that industrial relations would suffer from political, rather than industrial, reasons,' I told the MPs.

On a second visit to the Select Committee I said I thought there would be a coal shortage by 1975. In fact, events had proved that I was too optimistic: we had the shortage in 1970. One of the Conservative members of the Committee, Nicholas Ridley, later a Minister in the Conservative Government, asked me who would be responsible if by 1975 there was a shortage or surplus. My reply was: 'If there is a shortage of coal it is the Minister's responsibility and that of his predecessor. If there is a surplus of coal it is our responsibility. We think there will be a shortage.' And there was.

It is worth looking at how the forecasts of the Fuel Policy White Paper compare with the actual results. The first set of forecasts related to the year 1970 when, it was said, total inland energy demand would rise to 310 m. tons of coal equivalent. In fact, it turned out to be around 328 m. tons—so the forecast was a little matter of 18 m. tons of coal equivalent wrong. In coal this would mean the work of 36,000 miners for a year.

As I have said, nuclear power has fallen well behind expectations. The first stations, based on the Magnox type reactor, have suffered a substantial reduction in output because of unforeseen corrosion problems which may not be remedied during the whole of their working lives. In addition, there have been very big delays in the construction of the nuclear stations. Wylfa, the last of the Magnox series, due to come on stream in 1969, in fact is not yet commissioned. Dungeness B, the first of the Advanced Gas-cooled Reactor stations, originally due to be completed in 1971, is not now expected to operate before 1974.

These two factors of delay and corrosion meant that nuclear output in 1970 was 9 m. tons of coal equivalent—that is, 5 m. below the Fuel Policy forecast and even 1 m. below the actual output in 1969.

There is a similar story in the case of natural gas. Because of difficulties with the conversion programme, the use of natural gas has grown less rapidly than expected, and consumption in 1970 was about 16 m. tons of coal equivalent, compared with the White Paper forecast of 17 m.—nearly six per cent out.

Because of nuclear power's failure to come up to expectations, the whole of the country's additional fuel requirements have had to be met by coal and oil, with oil providing the greater part—at a time when it is rapidly going up in price.

The White Paper estimated oil use in 1970 at 125 m. tons of coal equivalent; the actual figure was 146 m. tons of coal equivalent.

Coal consumption was 154 m. tons, whereas the Fuel Policy gave us credit for only 146 m. tons unsubsidised and 152 m. tons with Government help to maintain the burn in power stations and gasworks. In fact, that help was not needed at all, so the White Paper figure was an under-estimate of no less than 8 m. tons. And, of course, coal consumption was less than it would have been because of its scarcity.

There is little doubt that if we had produced 200 m. tons of coal last year—my original 1960 target and the target of the Labour Party in opposition—it would all have been sold.

Inland consumption would have been greater because, for a start, it would not have been necessary to convert power stations to burn imported oil at the very time when its price was soaring. If the coal had been available, many industrialists would now, under the impact of fuel oil's increasing expense, be converting back to British coal. The steel industry would not have been forced to scour the world for scarce coking coal, paying famine prices far above ours. And, on top of all that, we would have been doing a very big export trade with the Continent, which has also shut too many pits and is hungry for coal, importing 25 m. tons a year.

In my last year we decided we would fix our own output objective. The aim now is to stabilise total output at around the present level of 150 m. tons a year. We formed our own opinions about demand and simply told the Government what we intended to do.

But the country would have been a lot better off if our advice had been listened to at the time the White Paper was being prepared. Instead of regarding coal as the residuary legatee, the planners should have started with coal. After all, it was the lowest risk fuel. It comes from under our own feet, not from the simmering Middle East. We know all there is to know about coal because we've been using it for hundreds of years, whereas nuclear power and North Sea gas are relatively untried and have let us down—especially the former. And coal is the only fuel available to us, apart from a little hydro-electricity, that imposes no burden at all on the balance of payments.

There was too much white-hot-technological-revolution thinking about the Labour Government's Fuel Policy. Nuclear power and North Sea gas were new and therefore, it was assumed, must automatically be better than coal, which was old.

Speaking of the White Paper and of an earlier and equally disastrous attempt to forecast energy supplies by a Committee under the Chairmanship of Lord Ridley, Dr Schumacher, the NCB economist, said: 'If, instead of getting all these experts together and using a computer-based model, the Government had asked the lady who cleans my office to do the job, she couldn't have been more wrong.'

Surely this country has at last learned the wisdom of maintaining a healthy coal industry. Because coal is the only fuel that requires absolutely no foreign exchange, it ought to have been the basis for the country's fuel policy, not the residuary legatee. The planners who prepared the White Paper were hypnotised by the prospect of cheap and abundant oil, a prospect that has gone for ever. British coal can compete on price with fuel oil and is still a much cheaper source of electricity than nuclear power.

Many people, particularly the country's Managing Director, thought I was being mischievous and even obstructive because of the clamour I set up at the time the White Paper was prepared. My demand, taken up by the unions in the industry and later by the House of Commons Select Committee, for an independent inquiry into nuclear power costs was said by Roy Mason, the Minister at the time, to be likely to frighten the life out of the coal industry because it would show that it had no hope of competing with nuclear power. That wasn't a clever remark even at the time it was made. And now it looks very foolish indeed.

It saddens me to think that the country has suffered because the advice that I and my colleagues (notably Schumacher) gave was ignored by the civil servants and the politicians. We were regarded as bloody nuisances at the time. We have the consolation of knowing that if we hadn't fought so stubbornly and made the whole question of fuel policy a public issue, the coal industry would have been smaller than it is today, and the country would have been in even greater jeopardy with its fuel supplies.

The Board, in their Annual Report for 1967/68, took the virtually unprecedented step of expressing public disagreement with Government policy by including a long section on the White Paper. We said that we disagreed with the long-term policy expressed in the document which, we added, 'would involve the displacement of such a large number of men as would be inconsistent with maintaining good industrial relations'.

In the Report we acknowledged that it was essential that there should be a long-term policy if the fuel industries were to deploy their resources of capital, technology and manpower in the way most beneficial to the national interest. And then we went on to set out our arguments why this particular attempt at fuel policy planning had been badly done. We were particularly critical of the size of the nuclear power programme and repeated, yet again, our demands for a full, independent inquiry into the technical and financial basis of nuclear power. In the following year's Report we returned to the attack and expressed doubts (since justified by events) about the growth of nuclear power and North Sea gas.

It was a grave thing to disagree in the pages of the Annual Report with the Government, but in our opinion they had made a very serious mistake and we were determined to have our objections on the record for everyone to see. For ourselves, we seriously over-estimated in 1967 the rise in productivity that we expected to get. This was the main reason for our disappointing financial results in some of the last years of my term. It meant that we were employing more men than we had expected to in order to raise the coal that the country needed. Productivity is closely related to morale, of course, and we under-estimated the effect on confidence of the very heavy pit-closure programme. It was perhaps expecting a lot of men that they should buckle down and keep the machines working longer during each shift, when they could see pits being closed all round them. Although this forecast did prove wrong, the warning I gave the Select Committee about the White Paper's effect on morale and productivity was proved to be well founded. The result was as I predicted.

Despairing of getting any thorough assessment of the coal industry's value or prospects from the Ministry and other Government departments, we decided to commission our own fuel policy study. We therefore invited the highly regarded Economist Intelligence Unit to do a completely objective study.

After a great deal of work over a period of about eighteen months, their report was published at the end of September 1968. What the EIU Report recommended basically was that the policy should be based on a middle of the road 'mix' of the primary fuels (coal, oil, natural gas, nuclear and hydro-electricity), which would require about 144 m. tons of coal in 1975, compared with the White Paper estimate of 120 m. tons in that year. The recommendations for the other fuels were that oil should contribute just about the same as under the White Paper, nuclear power and hydro-electricity together rather less, and natural and liquified natural gas considerably less.

The Report showed that to deviate widely from a coal output of around 144 m. tons in 1975 and 136 m. tons five years later, would be 'to buy the wrong package at the cost of ruining the economics of the whole coal industry'.

Naturally, we invited the Ministry of Power to consider the EIU Report and we sent them a copy about three months before its publication. Within hours of the document becoming public property the Ministry had issued a rejection of it. In stuffy language (one could almost visualise the civil servants looking down their noses when one read the words) the Ministry said that there were serious doubts about many of the figures and assumptions on which the estimates had been based. They ended loftily by saying that the Report did not introduce new information or evidence relevant to fuel policy.

Now, less than four years later, it is possible to see how wrong they were and how right were the Economist Intelligence Unit. But nothing we could do at the time had any effect at all on the Ministry. They had decided on 120 m. tons for 1975 and, though Ministers and Permanent Secretaries might come and Ministers and Permanent Secretaries might go, the decisions remained unchanged. Even today, after the necessity to import coal, which was the direct result of the Fuel Policy White Paper, that document still influences the attitude of Government departments.

The conclusion of the EIU Report that a 'Middle of the road' policy would give the best results was confirmed by another study I have already mentioned. This was carried out by the Brookings Institution of America. The report, prepared by a team of distinguished economists from Canada as well as America, covered fuel policy as part of a study of the British economy.

The authors were economists writing mainly for learned readers, so their language was somewhat technical. The report said: 'The latest plans probably exceed the optimal rate of substitution of, in effect, nuclear power for coal; they involve an exceptionally and unnecessarily capital-intensive method which will yield only modest gains in fuel economy and a relatively small relief of miners for other industries. Coal output should probably be cut back more slowly than is planned and the investment programme for electricity (especially nuclear generation) should be substantially reduced.'

The Report commented on all primary fuels, and was particularly critical of nuclear power. Later on there appeared this judgment: 'In electricity, innovation has probably been too rapid, in the first nuclear power programme after 1956 and probably now in the second generation of reactors.' It concluded: 'To summarise: "efficient" coal output during 1975 to 1980 is probably between 120 m. and 180 m. tons yearly. The exact level probably makes relatively little difference to fuel costs but a major difference to investment requirements during the 1970s. The net gains from accelerating shrinkage in the older fields in 1968 to 1970 are small or negative, compared with a more moderate rate. In any case, there does not appear to be a large pool of workers with low marginal productivity that needs to be squeezed or induced out of mining. Most of the shrinkage will occur anyway in the declining fields; speeding it up may impose high capital requirements and yield little in energy-cost reductions. If rapid shrinkage is to be done, the 1967 plans are well designed to accomplish it and absorb its effects. But the basic point—that it is probably not worth it—remains.'

The Ministry, fascinated by their new computer-based model, were still not impressed. I began to wonder just what cataclysmic event could shake them on this subject. The NCB, the Economist Intelligence Unit and the Brookings Report all proved far more accurate than the Ministry.

Richard Marsh had taken the Chairmen of the nationalised industries to Selsdon Park in 1967 for discussions before publishing his White Paper, as I have described. Roy Mason when he was Minister chose a different country retreat, Sunningdale, where it was expected that we should be more free from Press intrusions than we had been at the earlier weekend conference. The documents are still secret. After the conference it was announced that Mason would have another look at fuel policy with a special emphasis on trends in the coal industry. He may have had this 'other look', but the policy remained unchanged.

The Labour Government were most unsuccessful in their planning. George Brown's National Plan was a bold attempt to make a forward assessment of requirements in terms of resources and produce a proper order of priorities. It anticipated a growth in the national economy that was never achieved. The prices and incomes policy came to pieces in their hands. The lack of success was certainly not for the want of trying. But that particular plan and the Fuel Policy plan came to nothing. I wonder what the Managing Director thought about it all. Certainly these events and too much self-assurance cost him the election and prevented him from being the longest-reigning Prime Minister of this century.

Yet surely these two planning disasters should lead to some useful conclusions. After all, those who were physically engaged in the planning were people of high intellectual ability. There were lots of them and the salary bill was enormous. What went wrong? Simply this—that this sort of planning exercise should never be done by civil servants. They are neither trained in these arts nor are they long enough in their jobs to provide the study in depth based on experience. Ministers also suffer very often because they are short stayers. All the evidence goes to show that the long stayers are more successful. Harold is a case in point with three and a half years at the Board of Trade. Hugh Gaitskell, four years at the Ministry of Power in the Attlee Government, and Denis Healey, Minister of Defence in the last Labour administration for five and a half years, are others.

Planning is essential, in some detail for the near future, and in broad outline for the later years. But it must be a continuous process, not a 'one-off' job. And this means planning cannot be successful unless there is some sort of permanence about the planners.

I believe and have advocated that a possible solution would be to have a group of experts on energy matters who would constitute a Commission or Agency independent of direct Government control, but whose job it would be to advise on energy policy and undertake comprehensive studies. The Commission would be staffed by experts making their careers in fuel economics and in the specialist technologies. This Commission would be charged with doing the total sum that I have advocated so many times in the last ten years. Fuels would no longer be looked at in isolation. They would be planned in the national interest, and the capital available would be directed into those which would give the best return in national terms rather than in narrow advantage to one fuel or another, depending upon its fashionable appeal or the skill of its proponents in persuading Ministers and their advisers. At the end of the day fuel policy would remain a political decision but it would be based on proper advice.

The British Civil Service is composed of dedicated people who are both competent and efficient for the jobs that they are trained to do. But they are now frequently involved in running businesses for which they have had no training at all. Take for example the Ministry responsible for the nationalised fuel industries. Together these industries have an annual turnover of hundreds of millions of pounds. Their investment forms a very substantial part of the country's total use of capital resources. The industries individually are among the biggest business enterprises in the whole world. Private enterprise companies pale into insignificance beside them. Civil servants are quite unqualified either by experience or by temperament to be responsible for such huge undertakings.

The task of planning fuel policy is further complicated because investment in the case of most of the fuels takes a considerable time to mature. A new power station, for example, takes six years or so to build and should have a working life of thirty years. In the other direction, a decision to abandon a colliery practically always means losing for ever any reserves of coal that it still contains.

Departments are always timid to adopt the overall view, and instinctively think first and most of the time about their own departmental position. The National Coal Board have for very many years been required to submit for Ministry approval capital investment schemes over a certain size and those which will create new capacity. And yet in the Ministry to which we were responsible there were no mining engineers except for those in the Inspectorate responsible for safety. So how could the Ministry express a view on whether our investment schemes were soundly based or not? There was a similar lack of people technically qualified to comment on the schemes put forward by the other nationalised fuel industries. Then again, the promotion system of the Civil Service requires senior officials to be moved around fairly frequently. This means that by the time they have got a grasp of the problems involved in one department, they move on to another where they have to start afresh learning its business.

An independent energy board, providing advice to the Minister, and therefore preserving the system of Government and Parliamentary control, would be the answer to most of these difficulties. It would have a small but very competent staff of people who would be capable of doing the 'total sum', and of evaluating technically and from an investment point of view schemes put up by the different fuel industries. At the end of the day, fuel policy would still be a political decision but it would be based on much improved advice.

I argued the case for an independent energy board to the Select Committee several years ago, but when the Minister, Richard Marsh, was giving evidence to the same Committee, he could find nothing to say in favour of it. He claimed that the quality of the advice he was getting from his own civil servants could not be improved upon. The only snag is that all their forecasts turned out wrong. Every other agency that tackled the job gave better advice than they did.

I suppose that, simply because the idea was being put forward by the Chairman of the Coal Board, people suspected that it was intended solely to benefit the coal industry. But I was concerned as I always have been with the over-riding national interest. I have great admiration for Richard Marsh but I cannot help thinking that if he had had better independent advice from the kind of body I have been advocating, his 1967 Fuel Policy would not so quickly have tumbled in ruins.

Now he is running the railways, I wonder if in ten years' time he'll think differently. My guess is that he'll find it a change being on the other side of the counter. He is going to find, to quote John Davies, when he was the Director-General of the Confederation of British Industries, that 'you go into a room with Ministers, win all the arguments but you then have to leave the room and the decision is made without you'. Now Davies is himself the Minister responsible for coal and the other fuel industries.

The oil people in the United States saw what those in Europe and the United Kingdom responsible for fuel planning failed to see—that coal, and lots of it, would be required in the years to come. All the big oil companies are now suppliers of all kinds of energy. As L. F. McCollom, then Chairman and Chief Executive of the Continental Oil Company told me in his New York office: 'If a customer wants oil, or coal, or gas or nuclear energy we can give it to him. We don't have to turn away any buyer of energy.'

So most of the large coal companies in the USA are all part of oil firms. Continental Oil acquired the second biggest coal company in America, Consolidation Coal Company. Island Creek

Coal Company is now wholly owned by Occidental Petroleum Corporation, and Gulf Oil have acquired Pittsburgh and Midway Coal Mining. The giant Standard Oil of New Jersey has interests in coal deposits in five states through Humble Oil and Refining. In fact, no fewer than thirteen American oil companies now own and operate coal mines.

Continental were also interested in the long-term possibilities of producing oil from gas and one of their first moves after acquiring Consolidation Coal was to experiment with a process for this purpose. Their efforts were encouraged by the US Office of Coal Research, which expressed its belief that petrol could eventually be produced from coal at an attractive price.

The American coal industry had suffered a big rundown in the early 1950s, mainly because of competition from natural gas. However, by 1965 American coal output was back above 500 m. tons—an increase of about 38 per cent by comparison with 1954. The chief ingredients in this growth were the electricity and steel industries, the customers who have been mainly responsible for the strong demand in Britain and on the Continent recently.

Coal output remains at a high level in the United States. Despite all the errors of the Government planners it will in Britain too.

11

Safety and Health

The appalling number of 316 men were killed in the pits in 1960 and 1,553 were seriously injured. It is difficult to comprehend the misery, pain and economic hardship behind those stark figures.

Much hard work was being done to try to reduce accidents. Everybody was dissatisfied with the lack of progress and in 1961, my first year, the National Association of Colliery Managers adopted the theme 'Safety With Progress' for their annual conference, held that year in Edinburgh.

I listened to several thoughtful lectures as I sat waiting to give the winding-up address. It seemed to me that what was wanted was some impetus from the top. The whole approach needed to be dramatised and freshened, new skills brought to bear and new techniques developed for communicating the safety message. So when I got up to give the closing address, I announced that there was going to be a campaign of positive action. We were going to bring into that campaign all the determination and resourcefulness we possessed. The shameful accident problem must be licked.

The industry responded magnificently and swiftly to this strong lead.

The most important job was to get the safety engineering and training right. We made it clear that the authority of the safety staff would be strengthened and we set out to show how seriously top management regarded the campaign. The technical side of the attack was the one that was going to be foremost, but we also set out to exploit modern communications techniques to the limit. For this purpose we set up a National Safety Committee under the Chairmanship at first of Roy Glossop, Deputy Chairman of East Midlands Division. When Glossop retired after several years, he was succeeded by another distinguished mining engineer, John Brass, Regional Chairman for Yorkshire and the North Western coalfields. Their team included representatives of the unions, mining engineers and safety specialists from the coalfields and Headquarters, the Director of Public Relations, someone from the training side and one of HM Inspectors of Mines—indeed the Chief Inspector, Harry Stephenson, was a member for a good deal of the time.

While the campaign was building up, we got the results for 1961. It was the first year to be free of a major disaster (that is, an accident killing more than 10 men) since 1958. There had been a big improvement to a new low record of 234 deaths. The results were no cause for self-congratulation on anyone's part, but they did show that progress was possible: it was an encouraging start.

I inaugurated the drive by sending a personal plea to the home of every mineworker urging him to keep death and accidents out of the pits. I hoped the appeal would reach not only the 561,000 men that we had in the industry in those days, but also their wives, sweethearts and relatives. The important thing was, I argued, to refuse to take chances.

We started to have something about safety in every issue of *Coal News* and our other periodicals read by management and underofficials. In fact, all propaganda and communications techniques were intensively used—perhaps in a variety and on a scale never known in British industry before. Everyone acknowledged that safety engineering and proper training were the most important influences for good and safe behaviour in the pits. When it came to creating a climate in which all the practical work could flourish, freshness of approach was essential. Obviously, at this late stage in the industry's development, there was nothing really new that anyone could say about safety. What we had to do was to find different ways of saying it.

One method we adopted was to use mobile television units to visit the collieries and put on programmes that the men could watch at any one of five monitoring sets during the change of shift. They consisted of a mixture of films on safety taken from our own library sources, short films made at that particular colliery and covering local safety problems, and, best of all, interviews with local personalities. It was interesting to watch the men in the pit canteens when these programmes were being relayed, because it was obvious that as soon as they saw on the television the face of someone they knew personally that is what attracted the most attention. We were the first British industry, and possibly the first in the world, to use these techniques for safety purposes, and the mobile studios, of which we had five, are still an effective medium.

Another gimmick (and I am not ashamed to use the word) we employed for safety purposes was a machine called a Scopitone. The Charbonnages de France, the equivalent of the National Coal Board, had used this newly developed machine before we did. It was manufactured in their country. Essentially a juke box, it also showed pictures with the music. When a coin was inserted one got not only the popular tune of one's choice but also a film to go with it. What the French coal industry and we ourselves did was to insert before the film a short safety sequence. The Scopitones were set up in colliery canteens and attracted a lot of interest for a time. Like other gimmicks, we didn't over-use them because our policy all the time was to try to keep freshening the campaign by the introduction of new devices of this kind. Incidentally, we were able to buy our Scopitones much more cheaply than the French themselves. An importer in this country found himself stuck with a large number of them because they didn't prove as popular with the British youngsters as they were expected to be. I believe the reason was that it took the sponsors some time to fit the films to the music and by then the tune had become rather out of date. And, of course, youngsters are only interested in today's top twenty, not yesterday's. However, our people pressing their buttons in the pit canteens did not have to put any money in and we were able to acquire a large number of the machines cheaply.

Naturally, the unions were wholeheartedly involved in our efforts to improve safety. There is a Safety and Health Committee of the National Consultative Council and active joint committees which do equally good work at collieries, also as part of the consultative machinery. But not all of our gimmicks were enthusiastically received by everyone. We ran competitions whereby men who attended regularly and were free from accidents qualified for a draw in which there were some quite valuable prizes. Motor cars were particularly popular. But the leaders of the Notts miners thought that it was wrong to try to bribe men to safeguard themselves from accidents. They had missed the point, of course. What we were doing was simply to stimulate interest and thinking about safety.

Another controversial question raised in the safety campaigns was whether or not it was reasonable to use humour. The miners' leaders in one or two coalfields bitterly criticised some posters that were aimed at provoking a laugh, on the grounds that safety was a serious business.

I must emphasise that this was an isolated example of hostility by the unions to safety activities. Usually, of course, they were just as anxious as the management to see progress. Our people sensibly set out to try to find out from the men themselves what they thought. We commissioned some research into the effectiveness of all our safety propaganda activities and reached an interesting conclusion on this subject. It was found that the men *said* they preferred a serious approach, but from their answers to questions designed to measure their recollection of two posters, we established that slightly more of them remembered the funny poster. So we have gone on using humour as well as the earnest approach.

We have also in some of our films—they were produced by our own unit—used shock methods. Some of the material was far more horrific than any Hammer Films production. But we reckoned that any approach that was going to help put across the safety message was justified.

Again the research into the value of the different safety propaganda methods was about the first to be carried out in industry, certainly in this country, and probably anywhere in the world. It always seemed to me that we should try to measure the value of what we were doing, and safety was no exception. No one wanted to cut the budget for the safety campaigns, of course; we just wanted to be sure we were spending the money on the best things.

According to the research, films were regarded by the men themselves as being just about the most effective way of presenting safety material and they've been given high priority ever since. The films were extremely well done and their effect in reducing accidents confirms their value. For example, we made and issued a film called *Isolate and Check*. It showed the dangers of working on electrical equipment without switching off the current, and of ignoring basic safety rules covering this hazard. A report from the Barnsley Area said the film had been shown to every electrician. In rather more than twelve months after that showing, there was only one report of injury involving electricity. And that single exception caused only minor injury.

Incidentally, this was one of the films that used gruesome techniques: some of the injuries shown in it were ghastly.

Efficiency in getting coal was almost always reflected in a good safety record. The pits with the best productivity were nearly always also the pits with the fewest accidents. This kind of mirror effect was often true of bad results too. For example, pits with a high accident record underground often also had a high rate of accidents on the surface. Obviously, this could not be the result of difficult geology, because surface conditions cannot possibly be influenced by underground problems. Also, when we started our special safety efforts, the surface accident rate was much higher than the rate in other heavy industries. There could be no possible excuse for this and I chucked my weight around, demanding to know why it should be so. In finding explanations, people often found remedies.

After the promising 1961 results the accident figures stubbornly refused to yield further, despite all our efforts. In 1962, 255 men were killed and 250 in the following year.

Although 1964 showed a reduction in the number of fatal accidents to 192, progress was still much too slow. Harry Stephenson, Chief Inspector of Mines and Quarries, said in his Annual Report that although the industry had succeeded in some of its aims, more than half the accidents should never have happened. He made that statement, he said, after every allowance for human frailties. Regulations were being ignored and there was an almost incredible failure either to recognise obvious risks, or to provide adequate safeguards against hazards which were likely to arise.

It was clear to me and to all my colleagues that the pressure had to be kept up and intensified. New thinking was needed. We just had to lick the accident problem.

The fresh approach we adopted became known as the Fifty Pit Scheme. Every year we singled out the pits which, in their own Areas, had the worst accident records. Each campaign was the subject of a launching conference, to which came representatives of the management and of the unions at those pits. Mostly these launching conferences were held at Nottingham University during vacation time.

I remember that at the first, I sensed a slight holiday spirit. People were having a day off from the pits and were enjoying themselves a little too much for my liking.

So when I got up on the platform I really let rip. Deliberately, I set out to rile my audience. I said that usually when we invited people to leave the pits for a day it was to honour them or to congratulate them in some way. But nobody in that hall had won any medals. I went on: 'You are all here because you have failed. You have failed to get your accidents down to the level that has been achieved in the rest of the industry.' And I went on to say that because of that I regarded myself as having failed too. I made it clear that any of the pits that showed continued failure to improve their safety record would be included in the next year's list of fifty. Which would be something to be deeply ashamed of. And the management would be held accountable.

For a period of nine months the selected pits were the subject of the most intensive safety activity. The specialist engineers, some from other parts of the organisation, examined all the pits' practices and got the safety engineering right. Then all the publicity techniques—the mobile TV vans, film shows, exhibitions, posters, and every other means of keeping safety to the forefront—were employed. Above all, management and the union leaders knew that their results were being held in the fiercest limelight and so they gave greater attention to safety than ever before.

The results certainly justified everything that was done. In the case of each campaign the average improvement at the fifty pits was a good deal better than the average improvement for the industry as a whole. And even when the campaign had ended at a particular pit, everybody had been shown that improvements were possible and they did their best to avoid slipping back again.

However, from every campaign, there were still some pits whose names had to go forward to the next list. When they eventually won themselves the respectability of being omitted from the next list, I am sure that everybody at the pit was relieved. In more recent years, it has been a case of including some of the better pits in order to see if their record can be further improved. Almost always, it has been.

At about this time we started to appoint Safety Engineers at the pits instead of the former Safety Officers. The new men were required to have First-Class Certificates of Competency under the Act, meaning that they were qualified to manage a pit. I believe this raising of the status of the safety specialists had a salutary effect.

Also about the same time as the Fifty Pit Scheme was started Bill Sheppard, then Director-General of Production, who always gave safety foremost place in his work, and I decided we would single out for special investigation four pits with the very worst accident records in the country. Our list consisted of two in Yorkshire, one in West Midlands and one in South Wales.

I don't think I shall ever forget reading the reports written by Bill Wood, then Director of Safety, who led the investigations. Wood was a tough mining engineer who didn't mind upsetting people if that was the way to raise safety standards. But he also knew when a little encouragement was going to be the best way of getting results. He and his team crawled all over the pits. His reports were a dreadful indictment of management. In some cases fundamental weaknesses in the method of working were to blame; in others, elementary failures to do simple things like guarding machinery, or having adequate lighting were reported on. Too often Wood drew attention to untidiness, for example in the travelling roads, that not only represented a hazard but indicated a poor attitude to the job by management generally.

When we got the reports, we wrote very strong letters to the Divisional Chairmen concerned, telling them to take the action that was needed. Some managers were moved or reprimanded and we saw to it that this severe disciplinary action was well publicised so that everyone knew we meant business. At least one Division initiated similar special studies of their own, which was the sensible way to react.

But in one case senior management were obviously irritated by Wood's strictures and defended their pit's management. They even suggested Wood had been taken in by the plausibility of the workmen, but this showed how out of touch they were themselves. His determination to get at the facts simply could not be resisted: no one ever took him in. So back went another letter from me which was even stronger than the first.

It was not only the accident figures that improved at these pits. In most cases, productivity rose quite sharply, and the financial results also benefited. The simple fact that someone had shown interest in the pit was enough to produce a good response.

The relationship between safety and a pit's other performance was an interesting one. Many people, quite understandably, expressed the anxiety that rising productivity would mean more accidents. They were afraid that with the pressure on more sustained use of machines, safety would be neglected. However, we were easily able to counter this by showing that year after year the East Midlands coalfield, which was the most efficient of all, also had the best safety record. In fact, the safest Area of all, South Midlands, was the most productive Area in the safest, most productive Division; which proved that safety and efficiency went hand in hand.

As I have said, we were always on the lookout for new ways of keeping the interest in safety fresh and vigorous. Some of the Areas found that competitive quizzes between four-man teams caught the enthusiasm of large numbers of people. So we adopted the idea on a national basis. Working to rules similar to those of TV's *University Challenge*, the teams fought their way through eliminating rounds mostly staged in miners' clubs in front of big and intensely partisan audiences to Area and then Regional finals.

The National contests were staged at the Café Royal in London's West End and the atmosphere was stiff with tension. The quiz masters had a real job on their hands. David Jacobs and Hughie Green took a turn in the chair, and in 1971 the greatly respected Will Paynter came briefly out of retirement to tackle it. Each of the four teams would be accompanied, like a successful football side, by its own supporters. By this time, they would have been practising for months, so keen was their desire to win.

But the quizzes were valuable not just because the teams swotted away at their safety rules. It was estimated that a total of 20,000 people watched the contest each year and, as with the TV programmes, they would all be trying mentally to answer the questions.

In 1971 we ran for the first time a national competition exclusively for apprentices, bringing in a whole new class of contestants, and increasing further the total audience.

Despite the size and intensity of our efforts, it was not until 1966/67 that a real impact began to be made—both on the number of fatal accidents and on the rate. There followed a steady reduction in each of the next three years until the numbers fell to 82 in 1969/70, the first time they had been pushed down below a hundred. Alas, there was a set-back to 92 in my last year.

Right from the time of my Edinburgh speech I continued to identify myself with the question of safety. And I didn't shirk the consequences of our failures.

I introduced the practice of writing personal letters to the widows or mothers of all men killed in the pits. It was an unhappy task but I made sure that, because of my wretchedness over every such letter, I kept up the pressure for still further improvements. Those letters became fewer and fewer in number but there were still far too many of them, and I was conscious all the time of the grief that lay behind the statistics.

The greatest health hazard in coal-mining, as distinct from the accident risk, is pneumoconiosis—a disease caused by breathing dust. Unless detected early it can disable a man, making it difficult for him to breathe, and even kill him. But if it is diagnosed in its first stages, the sufferer can be transferred to dust-free jobs and the disease need not progress.

This has been a difficult problem to tackle. It is usually quite slow to develop and men who are certified this year as having the disease are the victims of the conditions that they worked in a good many years ago.

The average number of new cases diagnosed annually about ten years ago was 2,800; the latest figure was 624 in 1969. It is justifiable to take some satisfaction in this progress but, as with the accident figures, it gives no cause at all for complacency.

Great resources of professional and scientific skill have been brought to bear on the problem. The Board have a periodic X-ray scheme which allows their medical and research staffs to watch the prevalence and progress of the disease. Long-term research, which for the first time has shown the relationship between the amount of dust in the mine air and the development of the disease, has led to new dust standards, which were adopted at the end of March 1970.

For many years concentrations of dust had been measured at working places underground by the number of particles per cubic centimetre of air. It had been known for some time that it would be much better to base approved standards on the mass concentration of dust, but no instrument suitable for doing this existed. The Board's scientists and doctors, in consultation with the Medical Research Council and the Safety in Mines Research Establishment, set out to produce one, and the gravimetric sampler that resulted from their efforts has made it possible for the suppression of dust to be intensified in places where that is needed.

This British device has been specified for use in the United States, where legislation against 'black lung', as the disease is known there, is also based on gravimetric sampling standards. Furthermore, a United States Congressional Mission visited this country in 1969 to collect detailed information on our dust-suppression methods and standards. The delegation reported to Congressman John H. Dent, Chairman of the Sub-Committee on Coal Mine Health and Safety Legislation, that they were 'impressed by the thorough and extensive nature of the British pneumoconiosis study program initiated more than twenty years ago'. So our experience helped to influence the way the Americans tackled the problem too. I devoutly hope that this stubborn problem will at last begin to yield, thanks to this new weapon produced by British doctors and scientists.

In the spring of 1969 I opened the Institute of Occupational Medicine in Edinburgh. This centre was established by the Board as the result of initiative by their Chief Medical Officer, Dr John Rogan. Of course, it is particularly concerned with the needs of coal-mining, but can tackle work for other industries and really represents a national asset. A high proportion of the Institute's total effort is naturally concerned with pneumoconiosis.

A special team to counter the dust problem was set up in my last year at the Coal Board and a campaign using many of the techniques adopted for the general safety drive was started.

A lot of work still remains to be done before pneumoconiosis is banished from the pits, but one other former scourge, nystagmus, has virtually disappeared. Improvements in underground lighting and in the power of the lamp the miner wears on his own helmet have been responsible. The disease affected the sight of anyone suffering from it and also his sense of balance. Maybe this was one reason why the old miners often abbreviated the name of the disease to 'stagma' or 'the staggers'.

Another complaint which used to be much more widespread

than it is now is beat knee. This is something like housemaid's knee and in this case improvements in the design of knee pads worn by the miners have been mainly responsible for the progress that has been made. Periods of absence attributed to beat knee were about 10 for every 1,000 men employed a few years ago; now the rate is only about half that.

Many thousands of mineworkers give up their own time to be trained in first-aid work. In fact there are now nearly 17,000 men appointed for part-time first-aid duties underground. There is tremendous rivalry between the first-aid teams from the collieries and this culminates every year in a national competition in which the best teams produced by eliminating competitions in the coal-fields take part. I have rarely seen such tension develop in a contest, even at championship boxing bouts. Incidents are staged with realistically simulated injuries for the 'victims', the teams are tested against the clock, and so high is the feeling that the adjudicators need all the qualities of a Cup Final referee. The teams bring their own supporters by the coach-load, as in the safety quizzes.

There is also the incentive that the winners take part in a national competition against rivals from other industries. In 1969/70 the St John Ambulance Team from Markham Colliery in Derbyshire won not only the national title but also the European Competition at The Hague against rivals from six countries.

The industry's own National Competition has a parallel junior section and it is good to see the enthusiasm with which the youngsters, many with flowing hair, get down to their tests.

The effort to reduce health hazards and accidents must go on and on all the time. Every casualty is just one too many for the victim's wife and family. Nevertheless, our hard work and the ingenuity and determination of the many people involved was rewarded. In fact, the drop both in the number of accidents and in their rate per 100,000 manshifts worked was the biggest single improvement we registered in all our operations during my ten years.

This was the best result of all, both in human terms and in efficiency. And it must be remembered that these improvements were achieved at a time when the new machines and techniques were being introduced at an extremely rapid rate. Furthermore, productivity was also being pushed higher and higher. These factors might easily have produced a contrary trend in the accident figures. Very many people of widely differing professional skills contributed to this achievement. There were the mining and safety engineers; the people involved in training, who saw to it that good safety practices were instilled; the doctors, nurses and first-aid men; and the propagandists, whether they were film makers, *Coal News* staff or designers.

Then at the pits themselves, the progress resulted from determined attention on the part of management, underofficials and union leaders.

Mining, because of its nature, produces a tremendous spirit of loyalty towards the other members of the team. If a man is trapped by a fall of roof everybody, immediately, is involved in the job of getting him out. No one stops to think of his own danger. Certainly nobody waits to find out if the unlucky one is a particular friend or not. I believe continued channelling of that sort of team spirit into the effort to make Britain's pits still safer will go on, and that it will produce standards that could not even have been thought about ten years ago.

12

Aberfan

There was nothing unusual about the morning of Friday, 21 October 1966—nothing to warn me that this was to be a terrible and tragic day that would make the name of the little Welsh village of Aberfan an object of pity throughout the world before the day was out. I prepared early on as usual for the meeting of the Board, which was always held on a Friday, and we were in session when the first reports started to come in. Naturally the early accounts were sometimes contradictory, but it was clear that there had been an appalling catastrophe. The Board's then Chief Safety Engineer, W. A. Wood, was given the job of arranging the quickest possible transport to the scene for a party of three people. It consisted of W. V. Sheppard, then the Director-General of Production, Wood himself and a Press Officer. They joined a team of Coal Board people working under the menace of another slide from the tip. Miners from neighbouring collieries were quickly at work making a channel from the base of the tip to divert water away from the dangerous area. They were only a few of the 2,000 people who worked at the grim task of making the tips as safe as possible and of recovering the broken bodies, mainly of children.

A great mass of the tip complex, a liquified material twice the density of water, had swept down the mountainside destroying two farm cottages which lay in its path, killing the occupants, and on to engulf a school, where the younger children were just at their lessons. Of the 144 victims, 116 were children, mostly between the ages of seven and ten. Of these, 109 died inside Pantglas Junior School. Five teachers in that school were among the 28 adults who died. The number injured was 35, and 29 of those were also children.

An expert witness at the Tribunal later estimated that in the final slip about 140,000 cubic yards of rubbish came down to the lower slopes of the mountainside and into the village of Aberfan.

Two urgent tasks had to be tackled immediately—an attempted rescue operation and work to ensure that no more of the tip could slide. Nothing could stand in the way of these immensely difficult tasks. My own responsibility as Chairman of the Board was to ensure that all the Board's resources were made available to these ends. We had to get the technical men on the spot with the miners who came from the surrounding pits to augment the strength of the men from Merthyr Vale, the village colliery.

Colliery disasters are unfortunately part of the risks of mining, but as a result the Board's rescue teams are always efficient and at the ready. The rules, too, are clear. No one should go to the scene of a disaster unless he has a specific task to do.

The rule covers the Chairman like anyone else—in fact, even more so. The appearance of the layman at too early a stage inevitably distracts senior and essential people from the tasks upon which they should be exclusively concentrating. And so I waited for the signal to go; meanwhile I received regular reports on progress.

I was much criticised at the time for not visiting the scene immediately. All I can say is that, in similar circumstances, I would still stay away until my presence could be useful, and not a hindrance.

As it was, the number of people converging on Aberfan hampered the work that was going on. Thousands of tons of muck had to be loaded into lorries which had to make their way along the difficult valley roads to dispose of their contents and return for other loads. At the same time essential supplies of medical equipment, food supplies and earth-moving machinery had to get into the village. The Chief Constable of Merthyr Tydfil found his work greatly hampered by the traffic which choked the approach roads and kept ambulances and other urgent movement at a standstill.

My task, and indeed the best way I could serve the people afflicted by this tragedy, was to ensure that all the industry's resources were put at the disposal of the rescuers. This was effectively done among my own advisers and among the people who were able to summon these resources. The rest of the day was spent in obtaining the requirements of the rescue operation. W. V. Sheppard, now in Aberfan, indicated that he would let me know as soon as it was right for me to go.

That evening there was a dinner at the Battersea College of Technology as a prelude to the birth of the University of Surrey in Guildford the following day. As Chancellor of the new University, I had a key part to play and I knew that guests were already travelling to Guildford, many of them from overseas, for the ceremony. It was therefore too late to stop them and, after earnest discussion with the Vice-Chancellor, Dr Douglas Leggett, we decided that the ceremony must go on, but that I should probably have to leave before the events of the Saturday morning were concluded. The signal would be the call from Aberfan, and that would determine the time I should leave the ceremony.

The following day began in the solemnity of Guildford's new Cathedral and, for me, ended on the slopes of the tip at Aberfan. The inauguration of the new University began in the Cathedral. The Bishop of Guildford, George Reindorp, led the huge congregation in prayer for the victims of Aberfan and their relatives. He referred to the tragedy again in his address, and there can be no doubt that the hearts and minds and prayers of those at the birth of a new University were with the villagers of Aberfan. The message came through towards noon that I could now go to Aberfan.

I travelled by car to Gatwick Airport, where the Board's aircraft was waiting to fly me to Cardiff. My wife and Adrian Alderson, my Principal Private Secretary, were among those who travelled with me. As we drove towards the village we heard police appeals on the radio for volunteers to stay away from the area. A BBC appeal for people to go equipped with shovels made confusion worse confounded. The authorities were satisfied that they had enough manpower, and the newcomers were simply choking the approach roads. Police at road blocks were keeping people away unless they had the strongest reasons for continuing their journey. By now nearly thirty-six hours had passed since the tip had slid. An organisation had been created. Even so, the village was a dreadful sight.

I and my small party went straight to the colliery office, where I was given an up-to-the-minute report on the condition of the tips and the hazards that still had to be controlled. I then went to the broken school, where we stood for a while amid the coats of the children still on their pegs and the other heartbreaking evidence of young lives that were now over.

I had been previously advised that an alarm system was to be set up, so that adequate warning would be given in the event of the tip moving again.

It was now nearly midnight and I returned to the colliery, where the Press were waiting to see me. I started by saying that the Coal Board accepted full responsibility for what had happened.

Unfortunately, because it was by now so late and there were very few telephone links with the outside world, the only newspaper that reported this statement was the *Observer*. Their story was headlined: 'No hiding by the Coal Board, says Lord Robens', and began: 'The National Coal Board will not seek to hide behind any legal loophole or make any legal quibble about responsibility. Lord Robens, Chairman, said at a Press conference in Aberfan last night, "First we have to establish all the facts and when we have the facts, then the Coal Board will behave as a public corporation."' This is important in view of what took place later.

I spent the night at the home of Ted Lewis, the Board's Divisional Secretary in South Wales. He, Alderson, my wife and I discussed the disaster far into the night. There must have been some human error somewhere within the Board's organisation. However unusual the circumstances or scanty the technical knowledge, there had obviously been some failure on the part of experts to communicate with those responsible for disposing of pit waste. I knew that when I went back to the village the next morning I should be pressed to say who, or what, had been responsible. Obviously such questions would have to be answered eventually in the calmer atmosphere of an inquiry. In the meantime all local NCB officials who might have been involved were at that time still too deeply involved in the recovery operations to apply their minds to the events leading up to the disaster.

Next morning I returned to the village and was able to see in daylight more of the tips and their condition. I joined the Secretary of State for Wales, Cledwyn Hughes, at a press conference in Merthyr Tydfil Town Hall. There had been more heavy rain during the weekend and I said that this had caused some anxiety, so that the Coal Board's immediate responsibility was to ensure that the tip presented no further danger. I said that there would be an inquiry and that the Coal Board were determined to give all help that would be needed to get at the facts.

That Sunday morning was clear and bright, and when we walked up the tips I went to where the workmen had uncovered a spring in the mountainside which was producing water at the rate of about 10,000 gallons an hour.

Returning from the top of the tip, I was met on the mountainside by a television interviewer, and the following conversation was recorded:

INTERVIEWER: Lord Robens, could you tell us please whether you think there is any danger of further slides?

LORD ROBENS: I don't think that there is any possibility of a further slide. We shall get some bits of slips just up at that top where we are working, but we've got that well covered, and it doesn't mean that there is any danger to anybody at all. The big task so far has been to discover the source of water, other than rainwater and drainage, and then to get that water away. As you can see, there are two big jobs going on here by the Coal Board: first to get to the head of the spring that we discovered in the mountain, and to make sure that a proper waterway is made to bring that by a diversion track away from the village into the river; and then on this side we pick up surface and rainwater and smaller little streams, and get that diverted into the river. So, in point of fact, the big job is to get the water away. This, I believe has been done; and in fact I know it has been done.

INTERVIEWER: Does the Coal Board have any responsibility for this slide occurring? Do you think 'this' tip should have been continued to have been used?*

LORD ROBENS: Oh, I wouldn't have thought myself that anybody would know that there was a spring deep in the heart of a mountain, any more than I can tell you there is one under our feet where we are now. If you are asking me did any of my people on the spot know that there was this spring water, then the answer is, No—they couldn't possibly.

* Unfortunately some of the words on the recording are indistinct or even unintelligible. Where this is the case the words or phrases in question have been indicated by inverted commas.

INTERVIEWER: Was it possible to know that this tip was dangerous?

LORD ROBENS: It was impossible to know that there was a spring in the heart of this tip which was turning the centre of the mountain into sludge.

INTERVIEWER: The Coal Board—the local Coal Board people were quite happy with this tip before this particular '. . .'

LORD ROBENS: Well, you would have to ask them, I don't live here. I am the Chairman of the Coal Board, and these people were under the direction of the Board, and I am statutorily responsible for all that takes place. But I couldn't answer a question as to whether my local Manager was happy or whether he was unhappy. All I *do* know is that he did not know that there was a spring deep down under the mountain.

INTERVIEWER: Are you going to conduct any thorough examination into other tips facing a similar position to other villages?

LORD ROBENS: No, I am not going to do it—we've done it!

INTERVIEWER: Are you quite satisfied that there is no other village threatened in the way that Aberfan has been destroyed?

LORD ROBENS: I am satisfied that we shall require to do some work for local authorities, and perhaps for some private landowners who have tips; because, you see, we don't own all the tips here in South Wales. We only own the operational tips. And we already have our normal safety measures for taking care of tips. But I am bound to admit that the normal safety measures do not contain any means of knowing whether there is an underground spring. It may well be that new techniques that have been devised will enable us to put probing instruments in, to give us that information; and if so, arising out of this terrible disaster, further advances may '. . .'

INTERVIEWER: Finally, Lord Robens, how long do you think it is going to be before Aberfan returns to something like normal?

LORD ROBENS: Well, seeing what has been moved in the time that they have already been engaged, I would think that there's at least another week's work to move all that is required to be done, before a good cleaning-up operation can begin. So I think there is going to be work here for at least two weeks before you can say the place is tidy again and the work of reconstruction—whatever kind it may have to be—could begin.

INTERVIEWER: Lord Robens, thank you very much indeed.

The interview was never broadcast, as the recording included too much background noise, but it was to figure in the subsequent inquiry in a very significant way.

Five days after the disaster resolutions were passed in both Houses of Parliament declaring that it was expedient to establish a Tribunal. Cledwyn Hughes, as Secretary of State for Wales, announced that he had appointed Sir Herbert Edmund Davies, one of Her Majesty's Lords Justices of Appeal, who was born within five miles of Aberfan and therefore knew the area intimately; Harold Harding, a consulting engineer; and Vernon Lawrence, former clerk to the Monmouthshire County Council, to serve on the Tribunal, with Edmund Davies as Chairman. The Tribunal took evidence in public on seventy-six days, holding the first sittings at Merthyr Tydfil, and the later ones in Cardiff.

I was on a visit to the Yorkshire coalfield when the announcement of the appointment of Sir Edmund Davies to head the Tribunal was made. I saw him on television defining the Tribunal's objectives. He said that the Tribunal would be concerned with four broad questions: what exactly happened; why did it happen; need it have happened (was this a calamity which no reasonable human foresight could have prevented, or was it caused by blameworthy conduct by some persons or organisations); and what lessons were to be learnt from what happened at Aberfan?

However, as the Tribunal went on with its work, these simple objectives, which were really questions of fact, became clouded. It was soon obvious that what was required of the National Coal Board was a simple plea of 'guilty'. This we were all ready to give, but we were misled by the terms of reference of the Tribunal, which were framed to uncover facts rather than to apportion blame.

Immediately the Tribunal was appointed I offered to give evidence. The reply to my offer was that my attendance was unnecessary as this was a technical inquiry and I would therefore not be able to assist.

Naturally, it was galling to me as I read the daily transcripts of evidence to see criticisms of what was called my failure to give evidence. Subsequently the Tribunal criticised the lack of communication within the Board. There was certainly a lack of communication within the machinery of the Tribunal.

Mr Desmond Ackner, QC, as he was then, counsel representing the Residents' Association, was the chief critic. Obviously he had not been told that I had offered to appear nor been told my presence was unnecessary. As he went on criticising my non-appearance, it apparently did not occur to anyone to advise Mr Ackner of this fact.

On the tenth day, a Mr Phillip Brown was called and examined by Mr Ackner. The next day the examination of Mr Brown was continued by Mr Ackner, who asked: 'Had you shortly before the members of the Tribunal arrived heard a broadcast on television in which Lord Robens had been interviewed?'—Answer: 'Yes, sir.'

Mr Ackner went on: 'My lord, I have a copy of the transcript of what was said.'

Chairman: 'Yes.'

Mr Ackner: 'Do you remember the question being asked of Lord Robens: "Does the Coal Board have any responsibility for this slide occurring? Do you think this tip should have been continued to be used?" and Lord Robens's answer?' (Counsel then read the answers I have already quoted from the transcript of the interview.) 'Did you hear that being said?'

Mr Brown: 'Definitely, sir.'

Unfortunately Mr Ackner, having secured a copy of the television script, had not taken the elementary precaution of ensuring that that script had actually been broadcast. And thus Mr Brown was quite wrong in saying that he had heard that broadcast.

Right at the end of the hearings, in fact, on the 69th day, the

Tribunal Chairman said that they would not want it to be thought that I had had no chance to appear there. He went on to say that I had not expressed any desire to be heard. This was simply untrue, and I still find it odd that those who knew of my offer to appear did not inform him.

He added that the Tribunal would not have it argued that I had had no opportunity of meeting the criticisms levelled against me. Why did Mr Ackner not ask Edmund Davies to request my attendance? Was it a lack of communications like the one for which the Board was soundly condemned? Does it prove that in the best system of communications the weak link is the human being?

I was half a mile or so underground at Seafield Colliery in Fife when Edmund Davies issued his invitation. I was cleaning up in the pithead baths when the message reached me.

Nothing that I had to say helped the Tribunal in any way whatsoever. The decision in the first place that my presence would not help as this was essentially a technical inquiry was the correct one, and it continues to surprise me that those who had replied to my inquiry as to my attendance in the negative had permitted the criticisms to continue.

What did emerge with great clarity was that before the inquiry opened I had said: 'I am bound to admit that the normal safety measures do not contain any means of knowing whether there is an underground spring.' This was contained in the transcript of the television interview which Mr Ackner held in his hand on the eleventh day of the Tribunal hearing.

Dealing with this I said at the Tribunal:* 'I think it would have been apparent to anybody that what I was saying is that *the people on the site* did not know and it would have been impossible for *them* to know, *because we had failed as a Board to provide the necessary regulations* [sic] *to enable them to know. . . .'*

The Tribunal said that because my own interpretation of my own words was not the interpretation of other Coal Board witnesses nor of the Counsel for the National Coal Board, the Tribunal did not 'accept it for one moment'.

Counsel for the Tribunal said that statement meant that a vast amount of time had been unnecessarily spent on issues which

---

* Report of the Tribunal, p. 91.

were directed at establishing that very point. But the fault could scarcely be laid at my door, convenient though it may have been to have done so.

As the *Guardian* and *The Times* pointed out, the Tribunal could have called me at any time. If time had been wasted the fault was, as the *Guardian* put it, surely not on the Coal Board's side. *The Times* wanted to know why I had not been asked by the Treasury Solicitor to make a statement. 'Surely nobody can have imagined that that duty was rendered unnecessary by a brief appearance on television', the paper went on. *The Times* concluded that the delay was regrettable and unnecessary, but that the Tribunal itself could not hope to avoid its share of the blame.

The Coal Board's Counsel at the inquiry, Philip Wien, QC (as he then was), in his summing-up speech said: 'From the early days, it was clear that the Board would have to face the music.' He went on to point out that I had said two days after the disaster that the Board would not seek any legal loophole or quibble. (This was the statement I made at the midnight press conference the night I arrived at Aberfan, but reported only in the *Observer*.) Wien went on to say that my television interview, in which I said it was impossible to know about the spring in the heart of the tip, was the result of my personal observation and of information picked up at the site. The Treasury Solicitor and the Coal Board's Legal Adviser had taken the view that this evidence of mine would be of no assistance to the Tribunal. Yet I had, Wien added, been criticised for my absence from the inquiry, and the Board's conduct had been described as little short of a public scandal. 'This was not only a challenge that a lesser man than Lord Robens would have accepted with alacrity but had placed the Treasury Solicitor and the Board's Legal Adviser in an impossible position', Wien declared. He said in addition that it was not surprising that I had attended the Tribunal and quickly showed that I had spoken from personal observation only and that the television interview would have been of no assistance to the Tribunal.

(Early in 1971, I was attending a dinner at the Guildhall and was standing quietly in the ante-room reserved for those sitting at the top table, when I was approached by a gentleman whose face was familiar. It was Sir Harold Harding, the consulting engineer member of the Tribunal. 'I have come across to speak to you,' he said, 'because I think that you had a difficult time at the Tribunal, and I had a great sympathy for you.')

The Tribunal's Report was published on 3 August 1967. For the record I must repeat that on the day after the disaster I accepted full responsibility on behalf of the National Coal Board, the following day I said that our system of tip control did not provide for discovering underground springs, which was a fault on our part, and I offered myself as a witness immediately the Tribunal was set up.

The Tribunal said in the Report's introduction: 'As we shall hereafter seek to make clear, our strong and unanimous view is that the Aberfan disaster could and should have been prevented. We were not unmindful of the fact that strong words of calumny had been used before our Inquiry began. But the Report which follows tells not of wickedness but of ignorance, ineptitude and a failure in communications. Ignorance on the part of those charged at all levels with the siting, control and daily management of tips; bungling ineptitude on the part of those who had the duty of supervising and directing them; and failure on the part of those having knowledge of the factors which affect tip safety to communicate that knowledge and to see that it was applied.'

Later, the Report said: 'The *basic* cause of the Aberfan disaster was that for many years in the coal mining industry little or no attention has been paid to the siting, control, or management of spoil tips. It might be said that, for this reason, most of the men whose acts and omissions we have to consider have had, as it were, a bad upbringing. They have not been taught to be cautious, they were not made aware of any need for caution, they were left uninformed as to tell-tale signs on a tip which should have alerted them. Accordingly, if in the last analysis, any of them must be blamed individually for contributing to the disaster (and that, unhappily, is the conclusion we have been drawn to regarding some of the National Coal Board employees and staff who appeared before us), for all of them a strong "plea in mitigation" may be advanced.'

The Tribunal found that the blame was shared (in varying degrees) among the National Coal Board Headquarters, the South-Western Divisional Board, and certain individuals.

The Tribunal said that, having learnt that the tipping gang and its charge-hand had all been bitterly reviled in Aberfan and treated as pariahs, they must make it clear that they absolved them from all blame for the disaster. The men were untrained and were faced with a difficult situation.

The Report went on to name nine NCB officials who, the Tribunal said, were in varying degrees liable to censure.

The Board's view, expressed in our opening statement to the Tribunal, 'that the disaster was due to a coincidence of a set of geological factors, each of which is in itself not exceptional *but* which collectively created a particularly critical geological environment', was pretty curtly dismissed. In the Report the Tribunal commented: 'Nothing that the Tribunal heard or read throughout its 76 days of sitting tended even remotely to support such a conclusion.'

Work done by the Board's geologists and by contractors has proved that our original statement was, in fact, no exaggeration. There have been many investigations into the geological environment of pit tips since the disaster, as part of the work of ensuring that such an incident does not occur again. Of all the many site investigations carried out, not one has shown the presence of such a combination of factors as prevailed at Aberfan.

It is necessary here to use some geologists' expressions. Nowhere else had tipping taken place where the upper level of an impervious boulder clay blanket ended at a critical point and the largest spring yet encountered in all the investigations found its exit. This spring had a higher relative ground water level than any other encountered and was in a position affected by mining subsidence.

The expert conclusion is that this is an exceptional, if not unique, set of circumstances. Yet, as I have said, the Report almost contemptuously rejected our argument that this was the cause of the disaster.

There were a number of recommendations about tip safety in the future, including several which required new legislation. An important part of the Tribunal's Report related to tip slides that had occurred at Aberfan in 1944 and at other collieries in South Wales. It pointed out that there had been no obligation on the colliery owners to inform the Inspectorate, because no one working on these tips had been killed or injured. Even up to the time of Aberfan there was no requirement to report accidents on tips unless they caused death or serious bodily injury to someone employed at the mine. 'Accordingly, the Merthyr Vale Colliery Manager was not under any obligation to report even the appalling Aberfan disaster to the Inspectorate.'

With two small exceptions, no other country in the world had seen any need to make regulations about tips. The exceptions were in Germany, where the Mines Inspectorate issued an order in 1964 governing the construction of colliery spoil heaps, but not their siting, and in South Africa, where there was an Act which contained requirements covering slime dams. Apart from that, no legislators anywhere had foreseen any danger of the kind that emerged so dreadfully at Aberfan.

Quite fairly the Tribunal said: 'There was a total absence of tipping policy and this was the basic cause of the disaster. In this respect, however, the National Coal Board were following in the footsteps of their predecessors. They were not guided either by HM Inspectorate of Mines and Quarries or by legislation.'

The topic had not been mentioned either in the report of a Royal Commission on the Safety of Mines published in 1938 or in the Mines and Quarries Act of 1954, which covered safety in and at collieries, but contained nothing on the operation of tips. Some Coal Board premises are covered by Factories Acts, but they too ignore refuse tips.

Of the Inspectorate the Report said that in the absence of statutory provisions making them responsible for tip safety, it was perhaps not very surprising that they appeared to have given no thought to the subject. The Divisional Inspector of Mines, Mr Cyril Leigh, had told the Tribunal that his departmental instructions contained no reference to colliery tips and that such inspections as were made would relate to the mechanical equipment on a tip, its superficial appearance and the method of working, and that no member of his staff was competent either 'to be pretty certain or to be absolutely sure about the stability of tips'. The Tribunal commented that there was no record of any Inspector having visited the Aberfan tip complex for any purpose during the four years before the disaster, despite the fact that Mr Leigh freely stated that he regarded it as part of the duties of a Mines Inspector to pay regard to tip stability so as to protect the safety of men working upon it. 'No explanation for this apparent failure to inspect was forthcoming to the Tribunal', they added. The Report contrasted the way in which Inspectors were required to record punctiliously details of visits underground, with the arrangements for the inspection of the colliery surface.

The Tribunal showed commendable humanity in what they called the 'vastly disagreeable task of censure'. Introducing this section of their Report, they said: 'Whether or not named or adversely referred to in this Report, there must be many today with hearts made heavy and haunted by the thought that if only they had done this, that, or the other the disaster might have been averted. Of course, some will blame themselves needlessly, others, while blameworthy in some degree, will condemn themselves with excessive harshness; yet others must carry the heavy burden of knowing that their neglect played an unmistakable part in bringing about the tragedy.'

We had to consider what to do about the nine officials who were singled out for censure. The Board decided that those officials still in the Board's service ought not to continue in the jobs they were doing at the time of the disaster. In fact, out of the nine officials censured two had already left the Board's service, five had moved to other jobs in the Board's organisation, and only two remained in the posts they occupied at the time of the disaster.

Immediately after the Board had decided on their attitude to the officials, Bill Sheppard, Cyril Roberts (Board Member for Staff), and I met the seven men still on our payroll in my room at Hobart House. We discussed the Report with them and explained what the Board intended to do about their personal positions. The two men who were still in their old jobs at the time of the disaster met Sheppard and Roberts afterwards to discuss their transfer to other duties.

I am bound to say that, despite the humane way in which the Tribunal commented on the people found to be blameworthy, the inquiry raised again criticisms of quasi-judicial Tribunals which I and many other people have long felt.

A Tribunal of this sort can, in effect, while not imposing a penalty, censure a man as severely as if he had been tried in a Court of Law. But these Tribunals are not Courts of Law. There is no jury. The men whom the Aberfan Tribunal censured had broken no laws or regulations under the Coal Mines Act under which they could have been prosecuted. They were involved in an event that had the most terrible consequences, but society as a whole was also at fault in failing to foresee the dangers.

It should be the Tribunal's duty to establish facts. Then, if prosecutions of individuals are justified, they should be brought in the ordinary courts under proper procedures.

My own personal position had to be decided. I don't suppose anyone seriously expected that as Chairman of the Coal Board I could know the condition of every one of the several thousand tips, working and disused, for which the Coal Board were responsible. Yet, as a politician, I had always accepted the doctrine of Ministerial responsibility—that is, if something goes wrong in his department, however remote from his own personal knowledge, the Minister must be at fault.

After the Report had been published and discussed by my Board, I had no hesitation in offering my resignation to the Minister. I announced my decision to my colleagues at the end of the special Board Meeting where we decided on the action that needed to be taken on the Report. It was held in the same room where we had been sitting when the first news of the disaster came to us.

When I had finished speaking about my personal decision, Cecil King, then a part-time member of the Board, spoke up and said that everybody who knew me would realise that I could do no other than offer my resignation, adding: 'I am sure the Minister will not accept it.'

Here is the text of my letter to Richard Marsh:

My dear Minister,

You will readily appreciate that my mind has never been free from the tragedy of Aberfan since that harrowing day on 21st October last year. I have now studied the Report of the Tribunal in detail and accept their findings, particularly their unanimous view 'that the Aberfan disaster could and should have been prevented'. I am deeply conscious that the deficiencies further disclosed, which caused the dreadful deaths of 116 children and 28 adults, occurred under my stewardship. It may be that the doctrine of Ministerial responsibility does not strictly apply to me as Chairman of the National Coal Board, but I have spent all my life in public service and I feel bound by its rules. I have decided, therefore, that I must offer you my resignation.

In the days that followed, there was a great deal of mail from people who wanted to let me have their views. The newspaper columns were also full of letters for and against. In fact, most of the letters that came to me urged me to stay on.

The National Union of Mineworkers appealed to Richard Marsh not to accept my resignation. The letter, written on behalf of the Union's National Executive Committee, and signed by Will Paynter, said that my departure in the midst of the difficulties associated with the reconstruction of the industry would be to the detriment both of the nation and the industry. An opinion survey carried out by National Opinion Polls showed that three out of four of the people questioned thought that I should not leave my post.

I should have been less than human if I had not found these expressions of support, coupled in many cases with sympathy, very moving. It was no wish of mine that there should have been the highly publicised argument. I didn't ask for the public opinion poll to be held. A *Daily Mail* reporter asked the Coal Board's Press Office what the total of letters for and against my leaving were. He was given the figures and subsequently the *Daily Mail* criticised me for publishing them!

I was in New York early in September discussing, among other things, how this country could export more British mining equipment to the United States when I had Dick Marsh's message that my offer to resign had been rejected. It was just a month after it had been submitted.

The Minister said he understood my motives and recognised the deep concern felt by me and Members of the Board. 'Nevertheless', he wrote, 'I am quite clear that I should not accept your offer. When you told me of your decision you mentioned the doctrine of Ministerial responsibility, but I do not think such arguments can reasonably be applied to a Chairman of a nationalised industry, nor do I consider that the conclusions of the Tribunal are of a kind which call for your resignation.

'Nothing will ever erase from our minds the terrible tragedy of Aberfan, but we all have a duty to ensure that nothing like it can ever happen again, and I believe that in this, as in other aspects, your contribution continues to be important in the national interest.'

I should have accepted a different decision without complaint, but I was glad to agree to go on. There was still a lot I wanted to do in the industry.

Mr Wilson in his book put a completely different complexion on the matter of my resignation.* He writes: 'The question of resignation did not arise. In any event, the Minister reported to me that he understood Lord Robens was considering whether to resign—not over Aberfan—at the end of the Board's financial year, though he had not finally decided. We decided no action was called for.'

I heard differently. I was warned by two of his Cabinet colleagues that my resignation would be eagerly accepted. As he was involved in the decision, it seems surprising that his view of the non-acceptance of my resignation should be so different from that of the Minister of Power.

John Gordon, writing in the *Sunday Express,* made the charming suggestion that I should be offered a share of the disaster fund as compensation for any anxiety I might have suffered.

It is worth quoting his words in full: 'Lord Robens has had the stain of Aberfan wiped off him and his crown as King Coal repolished and put firmly back on his head by one of the most cynical whitewashing jobs in our generation.

'All that remains now is to offer him 10 per cent of the disaster fund as compensation for any anxiety he may have suffered from the Tribunal's judgment of him.'

The people of Aberfan showed me no resentment. I had many letters from bereaved mothers assuring me that the disaster was not my fault and when I went to meet them, as I did from time to time, there was always a cup of tea and a cake.

They were most anxious that the whole of the remainder of the tip should be removed, and I well understood their feelings. It seemed to me, however, that a scheme that included removing a good deal of the tip, landscaping, terracing, grassing and some tree planting on the lower slopes would have completely taken away the menacing impression. Any money available, I thought, should be spent on better housing in the village and providing better leisure-time amenities for the youngsters who remained and who would be born in the village as life went on. At the invitation of some of the mothers, 500 of whom had signed the

* Harold Wilson, *The Labour Government 1964–1970: A Personal Record.*

petition for complete removal, I went to Aberfan in June 1967 to hear their case. After listening very carefully I explained that we had a scheme prepared by consultants which would include terracing, grassing and planting at the disaster site and we would be, of course, willing to pay the cost of this. I explained that, if it were decided to remove the tips altogether, that would have to be a Government decision and it would cost about £3 m.

The best expert advice was that the tips could be made safe without complete removal. I think they believed that, but still they wanted this daily reminder of their loss obliterated.

I thought the whole matter was badly handled by the Secretary of State for Wales, Cledwyn Hughes. I was sure, knowing the people of Aberfan as I did by then, that the alternative scheme to make the tips safe and redevelop the mountain area so that its whole appearance would be completely changed and the scars of the disaster removed forever, had an outside chance of acceptance if it were properly explained to the whole population of the village. A very large model was therefore made to illustrate how the work could be done and I invited Hughes and his colleague Richard Marsh to come and see it at Coal Board Headquarters. They were extremely impressed by the idea and agreed with me that the villagers might well accept it.

I argued vigorously that a marquee should be erected in the village and the model put on show, first to the villagers' representatives and the various Committees, and then to the whole population, with the Secretary of State present. I also said that I would be willing to go down there for this purpose and emphasised the great importance of presenting the plan publicly and with great patience and clarity.

In the event my advice was ignored. The model was shown in Cardiff only and only to representatives of the official bodies in the village. A junior Minister was sent to Cardiff and I was not invited, although the whole idea was mine and the Coal Board paid for the model to be made. The Government finally went so far as to get the villagers to pay part of the cost of the work they wanted doing out of the disaster fund.

It was the meeting in Merthyr the Sunday morning after the catastrophe that finally convinced me how important it is when disaster strikes to have an organisation ready to deal with emergencies. The civil authorities were trying to control rescue operations from Merthyr, but their writ did not run as far as Aberfan, nor could it do so in the fever of digging and the strangulation of communications down the Merthyr Valley. Transportation in and out of the disaster zone was the key to the whole rescue operation. What was required at Aberfan was a single authority with the command and expertise to provide equipment and stores for the rescue operation, and to control movement within the disaster zone. This could only be done if there were adequate radio, telephone and dispatch facilities available. People engaged in different parts of the operation largely went their own way, and probably had to.

Only two organisations at that time and in that place could have provided most of what was needed. First there was the National Coal Board, with a base at Merthyr Vale Colliery and traditional lines of command with the mineworkers who were the main instrument in the attempted rescue operations. The Board also were able to provide lorries and earth-moving equipment, as well as all kinds of stores, shovels and so on. Furthermore they had vital communication links with the outside world.

The only other authority that had most of these requirements was the Army, and they were withheld until rather late in the operation. So that the rescue was largely carried out by a group of loosely held organisations of various police forces, Civil Defence, local authorities and ambulance organisations.

I am not, of course, arguing that improved organisation would have saved lives. In fact, no one who was buried by the slide was recovered alive. But in other disasters better and swifter emergency measures might achieve even that and could certainly result in the injured being removed and treated more swiftly.

In a speech to a British Association symposium on disasters at Dundee in August 1968, I spelled out a simple but effective form of organisation that would prevent some of the initial chaos and confusion at disasters, both at home and abroad. The ideas were based on our Aberfan experience.

There is a real need for provisions such as these and it surprises me that Governments have still not found it necessary to provide such facilities.

I urged that three vital elements of our mines disaster and rescue operations were essential in dealing with similar situations elsewhere. First, there must be a single unchallengeable directorate to control the operations. Next, this directorate must have the ability to control the movement of men, material, and messages within the disaster area. Furthermore, it should have the ability to secure relevant expertise without delay.

To deal with disasters away from collieries, I suggested that the Home Secretary and the Secretaries of State for Wales and Scotland should be vested with powers to nominate an authority quickly and invest it in turn with the necessary powers.

As to equipment, there should be in each region, permanently at the ready, a fleet of pantechnicons to provide: communications (telephone and short-wave radio), medical services, canteen, rest facilities, control, power (diesel generators), and a couple of Land Rovers. Suitable nominated staff from the police and health services would be so organised that they could arrive on the scene in hours. Arrangements would be made with the RAF for air transport where appropriate.

I said that three days after most disasters, effective arrangements have been set up. The purpose of my recommendations was to reduce that delay to a few hours. The cost would be trivial by comparison with the immense benefits this would bring.

But there was no Government move, nor has there been yet.

13

Environment and Enterprise

Coalmining over the last two centuries got itself the unenviable reputation of being the industry that did not care. The industry that did not care what effect it had on the environment; did not care how much it ravaged the countryside; and in the last analysis did not care what effect it had on men's lives. It was an industry that seemed to be crying out for a sense of public responsibility. I was very conscious of this and, as I hope I have shown, one of my primary objectives when I became Chairman of the Board was to make it clear, beyond doubt, that the National Coal Board did care and that management from the Chairman down were deeply concerned with the quality of men's lives. I knew that if we were to succeed, I would have to establish a real partnership in effort. I considered that one of my primary functions was to ensure that people working for the Coal Board had a pride in the mining industry and realised that the Board themselves were determined to do what they could to make sure that genuine job satisfactions were available to all those employed in it. I have described elsewhere in this book the ways in which this task was attempted and in a great measure accomplished.

But having established more meaningful consultation, having done one's best to ensure that communications were improved to such an extent that no one had any excuse for the old line of 'but nobody told me what was happening', there was still the apparently intractable problem of the effect that the industry had on the environment, on the external fabric of men's lives. The pit heaps were still there, as the world was so tragically reminded at the time of Aberfan. No one seemed concerned about derelict land—it was just a fact of life.

It must be remembered that at that time the current fashion for interest in the environment had not emerged. It would not be true to say that the ecologists, the landscape architects and prophets of doom were voices crying in the wilderness. They were shouting loudly at the door, but few people heard them over the sound of the television. The prevailing ideology was still 'where there's muck there's money'. This seemed to me to be totally wrong. Here was a nationalised industry, required by statute to be run in the 'public interest', which meant running it in ways significantly different from a private operation. As far as I could see, it was certainly not in the public interest to leave desolation where once there was activity, pit heaps where once there were grazing cattle, subsided land where once there was rolling pasture, blackened and derelict buildings where once there was thriving industry. I gave increasing attention to this problem, and a pretty difficult one it was too. Not only was there the inheritance of two hundred years of coal operations, but there was the continuous problem of dirt disposal. Every year on a production of 150 million tons of coal 60 or 70 m. tons of spoil are brought to the surface.

It is always too easy to ascribe blame, but not so simple to ascribe it fairly. Much of the environmental damage which was now the responsibility of the National Coal Board had in fact been created long before the Board ever existed. It had been caused by companies looking for the cheapest ways of carrying out their business. If there were any periods in the last hundred years in which most colliery companies felt that they were making sufficient profits to consider the wider public interest, I am not aware of them. Whenever the national interest was threatened, as in the two world wars, the Government of the day had to take control of the pits. The Board acquired on vesting date not only considerable assets, but also very substantial burdens. What is more, burdens which (unlike the assets) grew in size year by year.

The predecessors of the Board began the practice of tipping on mountain sides instead of elsewhere because the economics of coal production seemed to dictate that that was where the pit dirt must go. They were in a highly competitive situation and no one was prepared to 'throw money away' by undertaking the very expensive and totally unremunerative action of transporting pit dirt to a more suitable site just to avoid damaging the environment. It is doubtful whether they ever realised they were damaging the environment. In their eyes they were creating wealth both for themselves and the nation. Most of them were interested solely in the return on their investment. There was no one to make them do the 'total sum' and take into account the social costs. They, therefore, dumped pit dirt as near to the workings as possible to maximise profits. The concept of social costs would have been (as it still is to many industrialists) vague, woolly and academic, having no relation to 'real' life, instead of being fundamental to any civilised society.

Many people like myself believed social costs mattered. Someone had to start caring and it seemed to me it had to be the Board.

The problem was that, however conscious I was of the legacy of mining dereliction, the fact remained that the cost of doing something about it was considerable in relation to the Board's precarious financial position and I knew that if we acted unilaterally it would fall entirely on the diminishing number of coal buyers. I had no way of getting the funds needed from the Government, and there was the statutory duty to get the Board's finances to a break-even point which was not going to be easy, faced as we were with fierce price competition from other fuels, legitimate claims for improvements in miners' wages and conditions, and the paramount need for massive investment to mechanise coal production.

The main difficulty as I saw it was that if the Board were to attempt to clear up the industrial legacy of two hundred years, then the bill would lead to higher prices, a loss of business and more pit closures. This of course was totally unacceptable. The nation had benefited in the past from coal sold at prices which did not provide for intelligent dirt disposal arrangements. In my view it was up to the nation to provide the wherewithal to remedy this situation. Dereliction was by no means even mainly confined to the coal industry; it was a national problem calling for national solutions.

I therefore became an active advocate of action to do something about the way in which this country was allowing thousands of acres of land to become derelict every year. I became a member of the Duke of Edinburgh's 'Countryside in 1970' Standing Committee when it was formed in 1966 and at the Chartered Surveyors' Annual Conference, held at Nottingham in July 1966, I emphasised both the scale of the problem and also how relatively cheap it would be to tackle. I pointed out that, whereas there were about 100,000 acres of land which was derelict according to a very narrow Ministry of Housing definition (in commonsense terms the amount of derelict land was far greater than this), the hard core of 60,000 acres could be cleared up for about £40 m., which, spread over say a ten-year period, was hardly within the margin of error of the amounts being spent on aid and incentives to industry. I was not content just to try to arouse Government to action; I was aware (even though experience showed I greatly underestimated its extent) of the difficulty in mobilising public resources to attack this problem which had been with us so long that people took it for granted. I have often found that if one looks hard enough, two problems can be put together and stirred up to give a solution. This was my approach in this case and the second problem ingredient was opencast coal.

Anyone thinking of the Board's activities which affected the environment inevitably regarded our opencast coal operations in a very unfavourable light. It was widely thought that such activities resulted in the wholesale destruction of the countryside and so could only be accepted in times of crisis.

It was easy to understand how this attitude had arisen. As is so often the case, the reasons were mainly historical.

This technique of opencast mining can only be used where the coal is near the surface and is unencumbered by roads and buildings that have to remain. It involves stripping off the layers of strata which overlay the coal, baring the coal to the light of day and then extracting it. Where it can be done this method has obvious advantages and is certainly the cheapest way of winning coal. In the United States about a third of the total production of about 600 m. tons a year is won by this method, and it is extensively used throughout the world where there are vast areas of virgin coal seams in unpopulated areas. As these conditions manifestly did not apply in Britain, the technique was always thought inapplicable in this country. Crippling shortages of coal during the Second World War however made it imperative that any possible method of increasing coal supplies be fully considered. Sir Albert Braithwaite, MP, who was a director of a large civil engineering company, pressed this method upon the Government, and as a result opencast mining began in 1942 to meet the war emergency. This operation proved very successful, producing at peak nearly 15 m. tons in a single year. However, the general impression was that this was emergency action to meet a crisis situation and that once the shortage was over the country would revert exclusively to deep-mined methods of coal production.

The 'emergency' approach left lasting scars because of the way the operation was initially carried out. To enable opencasting to work, the necessary land was requisitioned under Defence Regulations, which the farmers disliked; although it was restored after completion of the site, there was neither the time nor the expertise available for adequate restoration, with the result that the land was impoverished. During my four years at the Ministry of Fuel and Power I was in charge of all the opencast operations. The responsibility for opencast coal was transferred to the Coal Board in 1952; since then a tremendous amount of attention has been paid to ensuring first-rate restoration of worked sites. Today it is now generally accepted that the restoration work often results in an improvement in the value of the land rather than the reverse.

As opencast coal was introduced to meet a coal shortage, it might well be asked why it should continue at a time of coal surplus. The short answer is economics. Since the Board took over the operation, opencast mining in Britain has been highly profitable. In nineteen years it has produced over 190 million tons of coal and provided profits of over a hundred million pounds to the Board's accounts. Without these profits, coal prices would have had to be raised and to that extent the general economic well-being of the country worsened. But there is more to it than that. The actual work of operating the sites is contracted out by the Board to civil engineering firms. As a result a whole 'opencast industry' has grown up to carry out work on sites, and to provide the highly specialised equipment needed. This is of a size not used in any other industry in Britain and opencasting thus provides the essential home market on which manufacturers of the giant machines can base their export operations. Something like £40 m. worth of the exports of major earth-moving equipment from Great Britain is entirely due to the existence of opencast mining in Britain.

But from an environment point of view the operations were not welcome. It occurred to me that opencasting could be thought of in terms other than the production of coal. It was a gigantic operation concerned with the shifting of about 150 million cubic yards of earth each year. This meant that landscapes could be created as well as destroyed. Form and feature could be given to an uninspiring setting; lakes could be created, derelict land could be reclaimed and, obviously, pit heaps could be buried, reshaped or reformed. Here I had my two problems offering a joint solution. Accordingly I instructed the Opencast Executive of the Board to see how far they could combine their coaling activities with the clearance of dereliction. The drive and enthusiasm which they put behind this new initiative proved very successful and the Board's reputation for restoration and reclamation soon became internationally known. In fact when I was in Washington on one occasion I spent some time with the then Minister of the Interior, Stewart Udall, on this very subject. They had had thousands of acres of land devastated by strip mining (their word for opencast) and the attempt to make these ugly scars presentable had produced pitiful results. Udall was intensely interested in our restoration work of which he had heard and, as a result of our conversation, he sent a team of expert advisers over to the United Kingdom to see it for themselves.

Ultimately legislation was framed to deal with the problem of restoration along the lines of our own British practices.

Following the tragedy of Aberfan, I wrote to the Prime Minister, on 31 October 1966, suggesting that in the aftermath of Aberfan the existence of a vast number of spoil heaps and other industrial dereliction was likely to be a major matter of concern. There should be a strong drive to clear up the hard core. I suggested a ten-year programme costing about £4 or £5 m. a year and offered the knowledge of the Opencast Executive of the National Coal Board for use on an agency basis by Local Authorities to get the work done.

About three weeks later Harold Wilson replied saying that he had asked his colleagues to 'look into my proposal'. From then on the dead hand of the bureaucracy of the Civil Service was firmly laid on this idea and, despite some public announcements of good intent, Harold Wilson's colleagues were still considering a national drive on dereliction when they went out of office four years later. During this period of intense inactivity the total acreage of derelict land in this country had increased by a further 12 per cent.

For four years I badgered successive Ministers of Local Government, Royal Commissions, Local Authorities, Planning Councils, Professional Associations, MPs and civil servants to get them to see that the clearance of dereliction was not just a question of aesthetics but of fundamental development economics. Tony Greenwood, who was then Minister of Housing and Local Government, was interested enough and I think might well have carried the idea forward except for an unhappy mischance.

We had discussed the whole question of dereliction over lunch and arrangements were made for a meeting to take the matter further. Unfortunately towards the day of the appointment Greenwood fell ill and I was asked by his Department whether I would like to postpone the interview or to discuss the ideas with his Parliamentary Secretary, Lord Kennet. This is where I made a mistake. I did not know Kennet, but understood that he was young, bright and an enthusiast for reform and I thought that he at any rate would push the project forward. But it was not to be: it seemed to me that he was sticking manfully to his civil service brief, which clearly was to press me about the tipping which the Board were doing off the Durham coast, and which was spoiling the beaches. But this was not the main purpose of the meeting and in any case it was being dealt with on the spot between the Area Director of the Board and the County Planning Office.

Try as I could it was impossible to cover the wider problem, as Kennet continued to return to the subject of the Durham tipping. After a while I gathered up my papers, told him I wasn't wasting any more time and left his office. Another opportunity lost, and for the Labour Government another lost initiative. Kennet, too, if he had but realised it, had missed a trick.

Derelict land is most prevalent, of course, in depressed and development areas, which are desperate for new industries and new employment. Governments have poured in thousands of millions of pounds trying to promote such developments. I tried to show them that a small proportion of the money that was being spent on incentives and grants, applied to the clearance of dereliction, would be far better spent.

It always seemed to me to be naïve in the extreme to expect to attract light, clean, growth industries such as electronics from the attractive South-East to sites in the shadow of dereliction. I was not alone in this opinion. In 1968 I gave evidence to the Committee set up by the Secretary of State for Economic Affairs, under the Chairmanship of Sir Joseph Hunt, to inquire into the problems of areas (other than development areas) where the rate of economic growth could give cause for concern.

In their report they accepted my broad thesis pointing out that progress in reclaiming dereliction had been disappointingly slow and that, at the then current rate, it could take thirty or more years before dereliction held to justify treatment could be eliminated. Moreover, this took no account of the annual additional increase in the area of dereliction still being created.

The Committee went on to say that the present organisation for reclaiming derelict land was slow and cumbersome. But three years and a 12 per cent increase in derelict land later, the same organisation still creaks on.

The Hunt Committee proposed that a derelict land reclamation agency should be set up as a means of assisting the Local Authorities in the execution of their reclamation schemes. They went on to explain that not all Local Authorities would need, or would necessarily wish to use, its services.

A further recommendation was that the Government should consider making greater use of the Coal Board's Opencast Executive as the nucleus of such an agency, which was a considerable tribute to the work that the Board had begun to do in this field.

They also proposed that a country-wide programme of reclamation should be drawn up reflecting national and economic priorities and that the Government should set firm targets for reclamation of all derelict land justifying treatment.

These all seemed eminently reasonable proposals and, at a cost of £7 m. a year on the Hunt estimates, they could hardly be described as expensive. However, the Hunt Committee report, like so many others, was buried without even the courtesy of a decent funeral service. No constructive counter arguments were put forward. The only firm reason for the rejection of these ideas that I was given by Ministers and civil servants was that to accept the Hunt recommendations would be to interfere with Local Authorities and this Local Authorities did not want. What a wonderful alibi for them to do nothing that was troublesome! Ministers could have altered this attitude but they didn't. As I have pointed out, this is a manifestly illogical reading of the report and, on a number of occasions following its publication, when I raised this very issue at meetings with Local Authorities and Local

Authority Associations I received full support for the Hunt proposals. This was another occasion on which Whitehall were determined to know best.

So yet again constructive proposals were defeated by the dead hand of what Max Nicholson has called 'The System'. It was another indication of the incapacity of Ministers to manage.

It has puzzled me greatly to try to understand the blind, apparently unreasoning, opposition of the Civil Service. From the comments I received from them, either directly or through the voice of whatever Minister was put up to speak to me, it was clear that they had little appreciation of the realities of the problem or the simplicity of the solution which was being offered to them.

I finally concluded that either they just could not bear to relinquish their power and authority over land dereliction in case the rapid progress that could undoubtedly be made under a Land Reclamation Agency showed just how pathetic the efforts had been in previous years, or perhaps they felt that there was some catch in my proposals which they had not spotted. My Principal Private Secretary was once asked by a senior official in the then Ministry of Housing and Local Government, 'What was Robens *really* after in making these proposals which had little to do with the Board?' The fathead thought that there must be some ulterior motive or I wouldn't bother myself about such trifles as clearing up the country with the assets which the State owned. He just could not accept that my proposals were genuinely made on the basis of an analysis of the problem. The idea that one should look outside the narrow confines of one's particular job to see how the wider interests of the community could be served, had apparently never occurred to him!

The illogicality of the Ministry's approach is even more mysterious when it is realised that as far back as 1966, apparently impressed by our efforts, the Ministry of Housing and Local Government approached the Board to see whether we could help to get more progress made in the reclamation of dereliction in Northumberland. We offered technical assistance by the Opencast Executive to the Councils, and subsequently the Ministry circulated this offer to all Local Authorities. It was only later when we said we were prepared to take the matter to its logical conclusion and offer the Executive as a basis for a land reclamation agency that the Ministry recoiled in virtuous horror.

I began to realise that some people may have thought that I was supporting the Agency because it might come into my sphere of influence, so I wrote to the Minister pointing out that this was not the case and that I would be happy to see a land reclamation agency formed on any basis so long as it had firm targets, funds and the necessary expertise. But all to no avail.

I have found it most depressing to read and hear all the fine words about the importance of the environment—now we even have a Department of it—when I know what opportunities have already been wasted. But the opportunities still remain. Now that I am gone from the Coal Board surely the last suspicion that I had a personal axe to grind must have disappeared. It is still not too late to take the Hunt Report down off the shelves and put into effect its main recommendations.

14

Mining Folk

It was quite impossible to do the job I did without becoming emotionally bound up with the industry. One felt very deeply about every colliery closure because shutting a pit had such a tremendous impact on the mining community. The pit brooded over the village; people said: 'She's finishing', or 'She's away again.' No one ever dreamed of asking who 'she' was; it was the pit from the moment a child was capable of comprehending.

Travelling underground, no man was passed without a word being exchanged, or a quick question or two about how 'she's away'. Where in any other industry has management got such a close contact with the workman? Men in pits work in small groups, some on their own, so that there is a great deal of personal initiative to be used as the job goes on. The miners share a common danger; whilst they don't talk about it, they are all the time conscious that it's there. This means a great deal of self-discipline and a watchful eye on the youngsters.

The pit is important to those who simply live in the colliery village, because their livelihood is also bound up with the pit. A pit employing 2,000 workmen means that with wives and children about 7,000 or 8,000 people are dependent on that weekly payroll. Then there are those who service the village, the butcher, the baker, the candlestick maker. All in all there are something like 10,000 folk depending upon the colliery wheels turning. This dependency upon 'she' draws a community close together because of the common interests and this creates a unity that many people outside the industry have found it difficult to understand. Attitudes take a long, long time to change. Even today, forty years later, it is possible to discern the attitude of the old colliery owners in the 1920s and the 1930s by the present spirit of the workmen.

I have had men in their twenties and early thirties complain about the way the workers were treated in pre-war years. Most of them were not even born, still less working at the time. Miners are ready to dig back into history at the drop of a hat, for everything about the pit has shaped their lives. Pits sunk a hundred years ago, and still working, even now dominate the men and the villages. The old customs die hard and the terminology is different not only from other industries, but also from that used in other coalfields.

It is the inter-dependence upon one another, plus the hardships of the pre-war years, that made the miners and their kinsfolk eternally loyal to each other and to their institutions. Whenever there is a pit disaster or someone trapped, there is never a lack of men eagerly volunteering to go in and rescue. Those who are not chosen from hundreds of volunteers to do the job are disappointed and hurt. None of them ever counts the cost in personal danger to themselves. That they could easily lose their own lives in the effort to rescue a trapped companion in a difficult place is of no consequence. What matters is that a 'marrer' (the Durham and Northumberland word for workmate) is in trouble, and nothing else. Loyalty runs deep and nowhere is it shown more clearly than in the attitude of the miner to his trade union.

This was exemplified in the stoppage of 1926; long after the end of the General Strike the miners struggled on alone, faithful to the conference decisions advanced by their leaders. Loyalty to leadership decisions continues strong to this day. The tragedy is that no real interest is taken in union affairs by the mass of the rank and file, and if the leadership is poor then this plays havoc with industrial relations. Poor leadership at local level, combined with the blind loyalty of men, has closed more than one pit. Men will walk out of pits or refuse to go down on the say-so of local leaders without knowing the facts of the case or understanding the issues involved.

It all comes back to the unity of the community. No mining village is without its willing workers to arrange a children's gala or an old folks' treat and they'll dig deeply into their pockets for the money to provide it. Every sport and pastime has its adherents. It is said that you could field a rugger team fit to play for Wales merely by shouting down the pit shafts. This is also true of first-class bowlers in Nottinghamshire and Derbyshire. All-in wrestlers, boxers, musicians in internationally known brass bands, first-aid and ambulance brigade men, athletics and show business. The Coal Board could field a first-class team in any activity without any difficulty. Miners injured and now paraplegics have won fame with their skill at archery. The mining community is almost self-contained and today the old halls remain with the inscription carved in stone showing they were the Mechanics' Institutes of yesterday. Education was eagerly sought after by those for whom the run-of-the-mill community life did not give full satisfaction. I remember Jim Hammond, the now-retired Lancashire miners' leader, telling me that he had painstakingly taught himself Greek in order to find an intellectual challenge that was otherwise absent from the surroundings of his youth. I sat with him when miners from other countries were present and heard him conduct a conversation in two languages. All self-taught.

Local Government, too, has its share of participants, and I suspect the Coal Board has on its books more mayors and chairmen of councils than any other employer in Britain.

It is doubtful also if there is any industry in Britain which has produced since the war so many outstanding trade union leaders as mining has. Jim Bowman of course demonstrated his ability both as a leader of the Union and as Chairman of the Coal Board. When he was Vice-President of the National Union of Mineworkers, he was largely responsible during the war for merging the old Mineworkers' Federation of Great Britain into the National Union of Mineworkers. He and Will Lawther, the miners' leader from Durham, had tremendous influence in the trade union and Labour movements outside the mining industry. It was a powerful combination that made itself felt not only at home but in the wider international trade union movement. Will Lawther was a memorable chairman of the TUC in the early days of nationalisation.

Sam Watson, who succeeded Will Lawther as leader in Durham, made an effective, indeed masterful, contribution to the NUM and the wider trade union movement. He was also a powerful member of the Labour Party executive. Sam led the Durham Miners for more than twenty-five years. There is no doubt that had he gone into Parliament, he could even have become Prime Minister. But he preferred to stay in Durham with his own people.

He refused to put a label upon himself, and didn't regard himself as being on the left, right or centre of the Party. Essentially a pragmatist, he was a vigorous opponent of Aneurin Bevan and a warm supporter of Hugh Gaitskell because he was convinced that the latter's practical approach to problems was much more likely to produce results more quickly than Bevan's ideological method.

If I were asked to classify Sam Watson I would say that he was a Christian Socialist. He started a socialist Sunday school thirty years ago and it remains to this day. Each year I have lectured there and since his death I have continued to go in a silent appreciation of a great man.

He was Hugh Gaitskell's closest friend at the time of some of the bitter rows in the Labour Party and I know that Hugh always had the deepest admiration for and gratitude to this man from a social background very different from his own.

Many people thought Sam ruthless and he was certainly very powerful inside the NUM. Often, although he didn't intervene in a particular debate at an annual conference, he was nevertheless controlling the whole business. Nothing much happened that Sam didn't want to happen when he was at the peak of his influence. He ran a kind of wholesome Tammany Hall.

I remember well the last annual conference he attended, which was at Skegness in 1962. Sam's final speech to this annual 'Parliament' of the Union was about training German troops in Britain. Naturally, this was the subject of a great row between the Left and the moderates whom he led. The debate was a passionate affair and Sam put the official Labour Party line, which was to accept German troops. He not only routed his opponents but insisted on calling in his high-pitched, squeaky voice for a card vote in order to humiliate them by showing to the world what little support they had.

That evening someone said: 'You're a hard man, Sam. Couldn't you let the Commies off the hook for once on your final appearance?' Sam replied: 'Why should I? They never showed any bloody charity to me.' And they didn't either.

Sam Watson was naturally always in his glory on Durham miners' gala day. He seemed to revel in the huge crowds, the brass bands, the good humour, and in acting as host to all the leading Labour Party politicians. Both Gaitskell and George Brown attended Sam's final gala and from one of the platforms on the racecourse ground, in the presence of about 200,000 people, praised Sam's work for the miners and the Labour movement. When the speeches were over and the crowd cheered him, this allegedly ruthless man wept unashamedly.

But he wasn't finished with the coal industry or politics. He became a part-time member of the Electrical Council and also a part-time member of the Coal Board, where his knowledge of the industry was extremely valuable.

Sam would never leave his native Durham although he was pressed many times to do so. Very early on in the first years of nationalisation he was asked to become a full-time member of the NCB but he refused, saying that he could make a better contribution to the coal industry from his position in Durham as miners' leader. Both Clem Attlee and Hugh Gaitskell did their best to persuade him to stand for Parliament. This would have been perfectly easy for Sam to do and he could have had the pick of the stonewall Labour seats in Durham. But all the pressures failed. This man was incorruptible; neither a bigger income nor greater power seduced him. Having made up his mind he could not be budged from Durham. And there he died, amongst his own, and they still talk about Sam Watson when they want to describe what is good about a man.

From South Wales came another great man with a political philosophy quite the opposite of Sam Watson's but with the same qualities of integrity and honesty. This was Will Paynter, for whom I have the profoundest admiration and respect.

Will Paynter was a great, sometimes amazing, trade union leader. There was a vivid example of this in June 1964. The Union and the Board were in the middle of wage negotiations when the annual gala of the Lancashire miners was held at St Helens.

We were both present—Paynter as a speaker, I as a guest. Because of dissatisfaction with the size of the Board's offer, there had been threats of one-day strikes and bans on overtime. That was the background against which the gala took place.

It was a wet day but still there was a good turnout. During lunch I sat next to Will, chatting about things in general, when he suddenly changed the conversation by saying: 'Alf, I've a difficult speech to make this afternoon. The crowd are not going to like it. Would you mind not sitting on the platform? I don't want anyone to think that your presence has anything to do with my attitude on this wages issue.' So after lunch, I quietly slipped away and my seat on the platform remained vacant.

He certainly had things to say that could easily have been misunderstood if I had been plainly present. I stood in the lee of a tent, partly sheltered from the rain, where I could watch and listen.

Only a man of deep conviction and complete integrity could have made that speech. His purpose was to lead men away from damaging the industry and away from the destructive influences that were challenging the official leadership. Paynter was still a Communist, which made his speech all the more remarkable, but his devotion to the Union and the men whose well-being was his responsibility, as always, came before his party affiliations.

He had the rare ability, found only in great speakers, of being able to combine logical argument and great clarity with immense passion and fire. On that wet day in St Helens he needed, and produced, all those qualities.

He told his hearers that he accepted that the Board's offer of 9s. 6d. a week on the minimum was the most the industry could afford. Industrial action should not be drifted into on the strength of popular slogans. Its advocates must show how it could succeed. A national strike placed at risk thousands of miners' jobs and could be justified only if some vital principle were at stake. No such principle existed in the present situation.

The coalfield ballot held a few weeks later showed a majority of more than two to one in favour of accepting the Board's offer. Once more, the members had ignored the militants, who didn't care whether they wrecked the industry and the Union. This time even South Wales showed a majority in favour of acceptance and only Scotland were against. The vote followed a national delegate conference which had recommended rejection. This was further proof that automatic mandates can never reflect the real views of the members. Only an individual secret ballot can do that.

This round of negotiations had been the bitterest since I joined the Coal Board. Paynter had saved the day and, in doing so, had saved many a miner his job. Harold Wilson was on the platform that day and congratulated Will Paynter on his courage.

Paynter's sense of probity led him to extraordinary lengths. It was always difficult to persuade him to have a 'working lunch' because—as he apparently flippantly but, I think, rather more fundamentally, said—he would never incur any suspicion of being beholden to the employer. On the other hand, he would cover long distances to tie up some formal agreement with me if he thought that by doing so he was helping to settle problems quickly. I well remember his agreeing to come up to Edinburgh to discuss some important outstanding matter when he had to be in Cardiff on Union business the next day. He arrived, late in the evening, looking somewhat travel-worn. I promptly offered him the use of the Board's Dove aircraft for the following day so as to reduce his travelling time. He jokingly replied that he would be happy to discuss how he could be quietly parachuted into one of the Welsh valleys, but as to publicly proclaiming his debt to the employer at Cardiff Airport, that was not for him!

When he retired at sixty-five from the Secretaryship of the Miners' Union, I argued in Government circles that this man's undoubted abilities and vast experience should not be lost to the nation. I had in mind part-time membership of a nationalised industry or membership of an inquiry committee.

No one was more pleased than I when Barbara Castle, who was then the Minister of Employment and Productivity, offered him membership of the newly established Commission for Industrial Relations. It was an excellent appointment.

He had, shortly after retiring from his post as General Secretary of the NUM, resigned from the Communist Party for reasons which, in common with many things that mattered to him, he kept to himself. As everybody who cared about industrial relations in this country hoped he would, he accepted Mrs Castle's offer. Inevitably sour suggestions were made that he had resigned from the Party in order to line his pocket. Will's response (in the *Morning Star*) is worth quoting: 'If it is to be the judgment of comrades with whom I have associated for the past forty years that I could engage in some unprincipled design of leaving the Party as a calculated step to a Government job, then they are as well rid of me, as I am of them!

'I hope, despite the present disagreement, to continue to serve the ideals which have guided my life over the past forty years.'

Subsequently he resigned from the Commission as he found the Conservative Government's policy on industrial relations unacceptable.

Paynter never hesitated to criticise his members if he thought by their actions they were harming the industry and jeopardising the jobs of workmates. This is not a popular thing for a trade union leader to do, of course, and nowadays is almost unknown.

In the spring of 1967 we were all especially worried by bad attendance in South Wales. Paynter went down there and address-ing an audience of Monmouthshire miners said that the record of absenteeism at many South Wales pits represented 'irrespon-sible, anti-social behaviour'. He pointed out that the average rate in South Wales was well over two percentage points higher than the average for Great Britain and this represented a loss of at least £2 m. a year in net income to the coalfield. He asked: 'How many more pits will face precipitated closure because the will to save them is undermined by this couldn't-care-less attitude?' For good measure he went on to demand: 'If this attitude to work is not changed, what really is likely to be the response of manufacturers to Government invitations to site industry here?'

The men accepted his criticisms because he had their respect and they knew all the time that his great ability was being used only in their own interests.

As a tribute to Paynter the NUM held their annual conference in his native South Wales in 1968. This was his last annual conference and it was at Swansea. He had been General Secretary during nine difficult years for his Union. The industry had been contracting, sometimes savagely, for virtually the whole of his spell. And yet his achievements on behalf of his members were tremendous. Without him I doubt whether my hopes of getting the industry off piece-work and eliminating the strikes that so often were caused by it could have been possible. We had our disagreements, naturally. I remember one blazing row at a meeting of the industry's National Consultative Council when he thought I was being harsh to a member of the Union's staff.

Paynter was at the height of his power when he retired and he was certainly badly needed to continue the work he had started. But the Union rule says officials must retire at sixty-five and that was that. If he had still been Secretary of the Union during 1969 and 1970, the strength of his leadership might have prevented the strikes in the autumn of those years and the self-inflicted wounds they caused the industry.

In addition to the Commission for Industrial Relations appoint-ment he was offered a seat on the Coal Board as a part-time member. To my regret, but not surprise, he refused it. In Will Paynter's book that would have involved going over to the other side.

Not long before he retired the NUM organised a demonstration in London against the Labour Government's fuel policy. The rather sad, orderly lines of men walked quietly from Hyde Park to Central Hall, Westminster, where one of the speakers was Will. It was a remarkable address. With devastating, bitter logic, he destroyed the Government's case. At one moment Will was quoting costs per unit sent out from nuclear power stations in thousandths of a penny. At the next, he was movingly emotional as he spoke of the Union's loyalty to Labour and how it had been betrayed.

The packed audience, his members from all the coalfields, were completely absorbed, even through the somewhat technical passages when he was concerned with comparative costs of the different fuels. Mistaking one of his eloquent passages for the end of his speech, they gave him a prolonged standing ovation. He was unable to finish what he had intended to say, but his members had said a lot to him.

Some miners' leaders were often apparently quite careless of the effect of their public statements on the industry and particularly on the confidence of their members. For example in March 1967, Sam Bullough, President of the Yorkshire Area of the NUM, in his address to the annual Council Meeting of his Area said that coal was almost a bankrupt industry. It is fair to say that Sam, whose health has not been good for many years, was not normally as gloomy as this. It must be remembered, too, that most of the NUM leaders were deeply loyal to the Labour Government and they were upset over broken promises made in Opposition by Labour Party leaders, and apprehensive about what was going to be done about the industry in the fuel policy statement then being prepared.

However, I was obviously concerned about the effect of Sam's remarks on the confidence of the people in the industry and I happened to be on a visit to Yorkshire the day after his comments were published. I declared: 'This industry is not bankrupt in terms of money, ideas or the quality of its people, their enthusiasm, or their determination to win out against the tremendous competition we are meeting these days.' I pointed out that productivity was rising rapidly, and in Yorkshire the increase was then much higher than the national average. Then I went on to list the failures of nuclear power and the problems of the oil industry as set out earlier in this book.

I acknowledged that Sam Bullough was a great supporter of the coal industry but I said that his remarks would be likely to drive away recruits and we wouldn't get sufficient people of the right quality to spend time at college or university training to be mining engineers. Unless we attracted young people to the industry it would have no future.

We had plenty of people outside coalmining ready to write us off; it really made me angry when our own men cried 'stinking fish'.

At a time when the militants were trying to stir up trouble for the Board in South Yorkshire, I made a visit to Cadeby Colliery in that Area. Norman Siddall, then the Board's Director-General of Production, also came to the meeting and after we had dealt with his subjects he left to take a turn round the car park and get some fresh air.

Outside, Siddall found a very small, very young apprentice and in his easy way started to talk to the boy and asked him what he was doing there. The youngster said that he'd heard the Chairman was here and he was going to wait until he came out to see him. Realising that I always enjoyed talking to youngsters who had joined the industry, Siddall encouraged the lad to stay and arranged for a press photographer who was around to get a picture of him. However, when the militant NUM officials came out one of them started shouting and bullying the little lad, wanting to know what he was doing there. When the boy told him, this lout shouted at the lad still more and drove him away. What an attitude! Especially in a nationalised industry. One would have thought that the Coal Board and its Chairman were going to treat this youngster in the way that children were exploited by the coal owners seventy years ago.

The Union has always argued that the job of Industrial Relations Member of the Coal Board should be filled by one of their own members. They had therefore been angry when Jim Crawford, former leader of the Boot and Shoe Operatives' Union, had been appointed to this post during Jim Bowman's Chairmanship, after having done about a year as a part-time member.

Imagine the outburst then when it became known that he was to be replaced by another non-mining trade unionist, Bill Webber of the Transport Salaried Staffs' Association. Bill had been a part-time member of the Coal Board, but the Union protested publicly and vigorously. Eventually they came to respect him and he, like Will Paynter, played a great part in the important and revolutionary early successes in our campaign to get rid of piece-work.

At the time of his appointment, the official leaders of the Union emphasised that they had no personal objections to Webber. Their protests were based on the principles involved in his appointment. However, some of the most disgruntled members of the Union Executive used to try to embarrass Webber at meetings as they had done Jim Crawford, by using pit terms which they didn't expect him to understand.

Bill came with me to 10 Downing Street for a meeting with the Prime Minister and Richard Marsh. Harold Wilson greeted Bill with: 'Sit down, Sir William.' Bill replied: 'But I'm not Sir William, Mr Prime Minister.'

'Oh, really,' said the Memory Man, 'I thought you were.' In fact Wilson was somewhat premature. Webber was given a knighthood in a later Honours List. He well deserved the distinction, for he had played a wonderful role in the trade union movement, both in his own Union and in the TUC.

Will Paynter and Bill Webber got on well together, although you could be forgiven if you thought differently after listening to an argument between the two. Webber could be quite waspish at times but he never broke Paynter's unfailing good humour. Both were from South Wales, both were leading trade unionists and both were life-long socialists. They believed passionately in nationalisation and the only difference between them was that Webber represented management and Paynter the workmen.

They were determined to do the very best for the men and for the industry. These were not incompatible aims. If the industry could be prevented from contracting, then there would be more jobs for the men and greater security. Will Paynter never forgot that and threw himself into the work of improving productivity and efficiency on every conceivable occasion.

The combination of the two Bills put millions of pounds into the Coal Board coffers and saved thousands of men their jobs.

Only people like myself working closely with these two men could ever evaluate their work for the industry and the nation.

The nature of coalmining has influenced the character of what the trade unionists regard as the other side.

Managing a colliery is not the easiest way of earning a living. The manager is responsible in law for the safety of the people who work in his pit. If he has one of the biggest collieries, he heads a business with £15 m. of assets, 2,000 men, £8 m. turnover and an annual profit of perhaps £1 m. The manager is responsible for all industrial relations matters at the pit, though he has been assisted in this big task by the recent introduction of assistant managers responsible for personnel matters. Before the negotiation of the agreements that got rid of piece-work at the coalface, many managers were spending too great a part of their time on wages negotiations. Only a small fraction of their day's work was spent doing the job for which they had been trained and for which they had been required by law to qualify over a long period—mining engineering.

The good manager is much more than just a boss. The fact that most mining communities are separated from the rest of society means they have had to develop their own social amenities. The manager is automatically expected to take a lead in this and to involve himself on umpteen committees, none of them directly involved in the work of the colliery, but all of which make a contribution to the standard of life of the people who work at the pits and their families.

It is a hard life but it breeds an exceptional kind of men. All the present seventeen Area Directors have been colliery managers. They dominate the industry, perhaps because of the long time they are required by law to spend in getting their qualifications.

As far as mining engineers are concerned, the Coal Board is virtually a monopoly employer. They therefore have a special responsibility for these members of their staff. People are sometimes promoted too far or too fast for their ability. Obviously this is the fault of senior management for picking the wrong people. In such cases it is better for the man himself, as well as for the organisation, if he is found a different job. A man who fails in line management may often be able to do a completely satisfactory job in some specialist field. One who has proved himself a perfectly good colliery manager may find that, when he gets a more senior post that takes him away from day-to-day colliery work, he is out of his element. It requires much more skill to fit people into jobs where they can be successful and have a satisfying career than it does to sack them. And the Coal Board has put the round pegs into round holes with considerable success.

Two mining engineers who reached the very top of the organisation and who exhibited all the best qualities that men develop in managing pits are Bill Sheppard, who is now Deputy Chairman, and Norman Siddall, who succeeded him as Director-General of Production at Headquarters in May 1967 and as Production Member of the Board four years after. Both of them had been highly successful General Managers of our North Derbyshire Area.

On my visits to the pits I was always accompanied, first by Sheppard as the chief production official, and later by Siddall. We must have travelled many thousands of miles in each other's company and been down hundreds of collieries. As a layman I got extremely good technical advice from both of them and because they were both so highly respected, any criticisms we made of the management on the spot were never resented. The manager knew, at least, that the most senior people in the Board would have an intimate understanding of his pit.

In Board meetings they always showed very good judgment even on subjects that were not their direct responsibility. Not for them the narrow, specialist approach. Anything that happened anywhere in the industry was of importance.

One of the great and memorable characters on the management side of the industry was Johnny Longden. He was the very successful General Manager of our South Barnsley Area, where, despite the thinness of the seams and the worked-out nature of some of them, he had nevertheless brought in mechanisation at a rapid rate and pushed productivity to levels that were almost unheard of in that part of the coalfield. I first met Longden during my early days with the Board when Bill Sales, the Divisional Chairman, organised a dinner in Yorkshire for all the Area General Managers. After the meal I was asked to speak, and I did—about the coal industry, its problems, possible policies and so on. When I had finished there was a dead silence. It was broken by the voice of Johnny Longden asking: 'What about brass?'

He was quite right, of course, because all our efforts had to be aimed at building an industry financially healthy so that it could keep its prices competitive, pay for all the new machines and equipment that were needed, and also offer decent wages and salaries to the people who made their careers in it.

Mining engineers are sometimes slow to start a discussion at conferences. Always at the Scarborough management gatherings I mentioned in Chapter Five, all eyes would automatically turn to Longden. He never failed to have ready a first, ice-breaking comment or question.

Longden had started in the pits at the age of fourteen as a pony driver. He has never lost the blunt, expressive speech of the miner. And with his powerful build and close-cropped hair he never ceased to look like one.

He kept bees and used to tell gullible visitors, as he spooned honey into his tea, that it was good for virility. When a party of French journalists were taken to his pit he further stimulated their interest in permissive Britain by telling them that contraceptives were available on the National Health.

Longden smoked a pipe. His favourite tobacco was that black plug which miners also buy to chew down the pit where they're not allowed to smoke. It caused quite an impression among strangers when they saw this rough-looking character get out of his chauffeur-driven Princess, take a chunk of plug out of his pocket, slice it up with a knife, stuff it into his pipe and engulf himself and them in vile blue smoke.

In 1962 he was sixty-three years old and had done quite enough to justify his taking things easily and looking forward to a richly earned, honourable retirement. I explained that nobody would feel any less respect for him if he refused my invitation, but I asked him if he would be willing to move to the troubled Doncaster Area, which represented a more wealthy part of the Yorkshire coalfield as far as reserves were concerned. It was also an Area that was stiff with mining as well as with industrial relations problems. Longden readily agreed to go and he also took with him from his old Area a team of his experts. They quickly made an impact there and Longden stayed on an extra year to see the overall output per manshift, which had been around 37 cwt. when he arrived, go up to 45 cwt. He was a gem of a man and it made me proud to have people like him working with me.

I suppose that during the last forty years I must have met and known most of the trade union leaders in this country. The earlier breed were often rugged individualists but, with a few exceptions, their successors have been much more stereotyped.

This was certainly not true of Jim Bullock, for fifteen years President and unchallenged leader of the British Association of Colliery Management. A Yorkshireman, he was outspoken to the point of offending people who did not know him and mistook his frank, vigorous approach for rudeness. After the unpromising start to our relationship, which I described in the first chapter, he and I got on marvellously well—mainly, I think, because we were completely outspoken with each other and never tried to play our cards close to our chests.

His word was never broken and if he was sometimes hard-hitting in discussion, I knew that he was only doing it in the interest of his members and of what he thought was right. There was no trace of any personal animosity in him, though he never left any room for misunderstanding his position on a particular issue.

Jim started as a boy pony driver in the pit and by sheer hard work and force of personality he qualified as a mining engineer and became a highly successful colliery manager. He is quite an orator and has also become a broadcaster and writer of distinction. When he retired a couple of years ago he bought the beautiful Georgian stables of Swillington Hall, near Leeds. The Hall itself had been the home of Sir Charles Lowther, a member of the Lonsdale family, who drew their 6d. a ton royalty on all the coal produced from under their land. (Incidentally, mining royalties were nationalised by a Conservative Government as long ago as 1938.) The stables were derelict when Jim bought them, but over the years he has lovingly and carefully restored them and made himself a fine home.

I reckon he gets a deal of quiet satisfaction as he, the ex-pit boy, ruminatively pulls on his pipe and thinks of the aristocratic predecessors who first created the graceful surroundings in which he now lives, enjoying his richly earned retirement.

During a series of conferences with colliery consultative committee members in each coalfield, I had to go to America. One such conference was arranged to take place in a Wolverhampton cinema the morning after I was due to arrive back by air from the United States. Because of fog at London Airport, my flight was diverted to Scotland. The local Coal Board arranged for a car to take us down to Wolverhampton in time for the conference. We covered most of the distance the night before and decided to get a few hours' sleep in an hotel. Next morning, the fog made it difficult for us to get on. The starting time of the meeting, to be attended by men from collieries all over the Midlands, came and went. John Brass, the Divisional Chairman concerned, was at the cinema and announced that the start of the programme would be delayed. He had never seen one of these presentations before, so he got the people who ran them to tell him what the routine was. He then went out on to the stage and explained that I was on my way but that he had decided to start. He introduced the film; by the time it had finished, I and my party still had not arrived. So Brass went out on to the stage again and commented on the points in the film that we particularly wanted to emphasise, in the way he had been told I had done at previous showings of the film.

However, the crucial part of these presentations was when I dealt with questions sent in by the consultative committees. Brass was naturally reluctant to take over this bit of the programme and he and the stage managers were just thinking about asking the cinema manager to put on a few Donald Duck cartoons when I and my colleagues arrived. I took off my coat as I walked down the side aisle of the cinema, listening to a report on what stage in the proceedings had been reached. Then I went straight up on the platform and started dealing with the questions as they were drawn out of a drum.

It was typical of John Brass, another mining engineer, that he had not flapped but had got on with the job calmly and effectively. He is now a member of the National Board.

When we started our policy of concentrating output on a smaller number of rapidly advancing coalfaces, it was necessary to convince the industry that it really was possible to extract coal at such high rates. We therefore began by selecting a few faces which we intended should be practical demonstrations of what was possible. They were designated 'Spearhead Faces'.

One of these was at Sutton Manor colliery near St Helens and I remember going to a conference with representatives of the National Union of Mineworkers when we were explaining this policy. Ken Moses, the only Jewish colliery manager I ever came across, was so eloquent about the speed with which his face was moving, that there was a lot of banter about the risks of letting down Chester Cathedral by surface subsidence. The Cathedral is only fifteen miles from the colliery! At the time there was a big argument about whether Lancashire would supply the coal to the new Fiddler's Ferry power station or whether it should come from Yorkshire. Moses said that if the Central Electricity Generating Board insisted on importing this 'foreign' coal, he could always run a face rapidly under the power station and tip the boilers on end.

One of the people who, during my Coal Board time, brought much credit to the industry was surprisingly enough a girl. She was Dorothy Hyman, Britain's champion sprinter, who was the daughter of a miner and worked as a tracer in our Barnsley office. We did our best to help with training facilities in the somewhat isolated village where Dorothy lived and, whenever she came home from one of her triumphant Games meetings abroad, we gave her a little memento from the industry to go with the official silver and gold medals.

When she was at the height of her career Dorothy's father died and she helped to support her younger brothers and sisters. She was a shy lass, but obviously determined always to do what was expected of her and to be a worthy representative of her country and of her own people. Her return from the European Games in Belgrade, in 1962, where she won gold, silver and bronze medals, coincided with the annual staff flower show at Hobart House. I arranged for Dorothy to be met at London Airport and to be brought to the show, where I presented her, on behalf of the industry, with a bracelet chosen by my wife. Although she obviously didn't relish making a speech she knew that one was expected of her and with her characteristic determination she made a first-class job of her reply.

I must end this chapter with stories about some mining folk whose names I never knew. They are typical of many incidents that represented the close family feeling of the industry—even towards the Chairman.

Earlier, I mentioned an eventful visit to Newstead Colliery where I was met by two demonstrations, one of women who were complaining about the houses they lived in. There was one particularly outspoken representative of the 'gentle' sex. She looked like the grandma in Giles's cartoons. Some of her friends, taken aback by her bluntness to me, started to try to quieten her down. 'You want to mind what you're saying, that's Lord Robens', they said. 'I know it is', said Grandma. 'He's one of us though. He's Labour, like us, isn't he? We put him where he is.'

She was right, too. Politics opened up a new world for me as it has done for so many others.

I've described how the people of Aberfan and some of the local organisations invited me to go back to the district on a number of occasions after the disaster.

On one visit I was to be the main speaker at a dinner in the Town Hall at Merthyr Tydfil, go down Merthyr Vale Colliery (the one that made the Aberfan tips) next morning, and then meet the Residents' Association that afternoon. The weather was bad and the Dove was unable to take off, so I travelled by rail. The train was late. There had been a misunderstanding about the arrangements for a car to take me from Cardiff Station to Merthyr Tydfil and the local Coal Board PRO came to the rescue and drove me and the other members of my party to Merthyr over some very icy roads.

When we got out of the car outside the Town Hall a lad of about sixteen, big for his years, came striding towards me, a great smile on his face and hand outstretched for about ten yards of his walk to shake mine. He greeted me in a lovely musical Welsh voice: 'Welcome to Merthyr, Lord Robens.'

He must have been waiting in freezing cold for hours because, as I have said, I was very late. It turned out that he was an apprentice working at Merthyr Vale Colliery. We chatted for a few minutes and then I went in to the meeting and off he went into the night calling out: 'See you down the pit tomorrow then.'

I was very moved, of course. That youngster's attitude typifies all that I loved about the people of the industry. He obviously thought that the Chairman of the Board would appreciate being welcomed by somebody from the industry. And I certainly did.

And finally, out of hundreds of stories about the men in the industry, let me tell one that may be appreciated by those, like me, getting close to retirement age.

I was in the Club at a Yorkshire colliery village having a pint of beer with some old pitmen long since retired. They had been playing dominoes but stopped and made a place for me for a few minutes' chat. One of the old men was eighty-eight but looked as fit as a fiddle. His fresh, though lined, face, with the blue streaks of the miners' scars on his forehead, was illuminated by the merriest twinkling blue eyes that I have ever seen. He was the oldest of those present, enjoyed his pipe and his pint and his early morning walks in the surrounding country.

'How do you manage to keep so young?' I said. 'Weel, meester, it's like this,' he said, thoughtfully drawing on his pipe and then letting a great cloud of smoke rise to the ceiling before he went on, 'never put your slippers on, while you can still put on your boots.' It's a thought!

15

Report to the Shareholders

The State-owned industries are vast enterprises far exceeding the biggest in the private sector. Their relationship with Government, through Ministers and civil servants, makes them completely different from other businesses. Managing these industries is an art that has not yet been mastered, but is developing year by year. Any success that coal has had in its internal management has been due to the high quality of the men from many different professions who have been involved. One noticeable feature has been that since my own appointment at no time has it been necessary to go outside the industry for the men capable of managing its main activity. They were all there when I joined the Board, eager to accept the changes that came about—and positively welcoming them. They have all been involved in the changes, indeed they contributed vigorously to the thinking that produced them. My job was made comparatively easy by the quality of the men then lying at second and third tier down, but full of enthusiasm and waiting to go.

The success of the industry cannot be measured just in terms of productivity or profit and loss, but much more in the handling of a vast army of men facing redundancy and unemployment. All that we could do at Hobart House was to produce the plans; the real hard work lay at the pits, where local management had to break the news of pit closures across the table, followed by mass meetings of men who earned their living at the pit.

This was the really hard and difficult task and I have nothing but admiration for the way colliery and Area management carried through the operation, the size of which has never been faced anywhere in the world before.

It was the same skilful local management based on this personal approach at pit level which was the key to improved safety. Fewer men were killed and the rate based on manshifts worked was also reduced.

This was the factor that counted above all else. Whether in connection with safety or closure, the men knew that management cared. It was this knowledge on the part of the men that enabled the closure programme to proceed without strife.

My ten years have seen more than 400 pits closed and 64,200 men made redundant. Even if the closures had been spread evenly over the ten years and evenly over all the coalfields, there could easily have been strikes on a big scale. In fact, we never had more than one or two local stoppages about closures. And the closures were not evenly spaced out. At one time we were shutting an average of three pits every fortnight after the 1967 Fuel Policy was published by the Government, and areas like Durham had much more than a proportional share of the total.

It is scarcely possible to imagine the upheaval and the impact on people of this gigantic rundown. Whenever we had to take a decision to close a pit, I was always conscious of the effect, not only upon the miners and their own families, but on the whole community—local shopkeepers, people employed in the service industries, and so on. We were credited with having done a tremendous job in transferring men from closing pits to continuing ones in their own coalfield and also in coalfields perhaps a hundred miles away. We certainly worked hard at the problems and did everything we possibly could. But, at the end of the day, there were always some casualties.

Some of the saddest cases were men who had been injured in the pits and were able only to do light jobs. When the pit shut it was not possible to find light work for all the men in this position, for every existing pit already had its quota. When men and their families moved away to jobs in other parts of the country, there was the upheaval to face in leaving villages where their families had probably lived for over a hundred years. Friends, relatives and neighbours all had to be left behind. The impact on the villages based on the pits that had shut was bad. The young, active members of the community went away, leaving the old and disabled.

It was never possible to think of a closure only in terms of its effect on the balance sheet. I felt particularly sorry for the men in the North-East. Year after year they had to endure a regular succession of closures and they did this with dignity and without self-pity. But I remember on one visit to Durham being very moved by an old trade unionist who asked: 'When is it going to stop?'

Of course many of the pits died from natural causes, as it were. Their reserves were exhausted. Nothing anyone could do would change that.

But I think we all regarded closures which were due to economic reasons as a defeat for the industry. Many were caused by outside forces. The gas industry, for example, switched into other processes for making gas without the use of coal. The Coal Board were left to grapple with the social problems that this caused.

Obviously some of the pits closed for economic reasons were making such enormous losses that there was no alternative to closure. Good men and management were flogging themselves into the ground to try to achieve the impossible, when conditions meant that the pit never could be profitable. In such cases the skill and experience of the people involved were more profitably used in the long-life pits to which many of them transferred. Again, one regretted that such closures were necessary, but they were also unavoidable.

Of the remainder there were many that could, and should, have been saved. I deeply resented the attitude of many Labour MPs who claimed that it was the Coal Board who were shutting pits, not the Labour Government. These people had the gall to make such claims when their own Government's Fuel Policy said bluntly that coal output was going to come down very sharply.

Even before the 1967 Fuel Policy there were other Government measures, some imposed by Conservative Governments, some by Labour, which forced the pace on closures. For example, there was a limit put on undistributed stocks of coal. This meant that output not then needed had to be lopped off. The Conservative Government, with their White Paper on the Financial Obligations of the Nationalised Industries, also made closures inevitable. These obligations were continued under the Labour Government, so there was little to choose between them.

It is undeniable that we closed too many pits and that those unnecessary closures were the result of errors in fuel policy.

Again, the 'total sum' approach for which we argued repeatedly ought to have been applied. The State paid out many millions of pounds in redundancy payments and other social benefits for men thrown out of work. The Labour Government's Coal

Industry Act of 1967 made possible the payment of pensions to men of fifty-five. This provision was intended to, and did, prevent financial hardship, but what of the psychological effect on a man who is told at the age of fifty-five that he is no longer needed? All the money spent on protecting men from the harshness of unemployment was obviously needed. But it was also unproductive. No Government calculations were made of the cost to the community, through the Exchequer, of keeping men at work producing real wealth in the form of coal, even if there was a temporary surplus. George Brown's National Plan, which after about eighteen months was not worth the paper it was printed on, talked grandly of getting men out of mining and agriculture and into other industries. Many of these men, who in their former occupations had after all been producing real wealth, were offered jobs like making one-armed bandits, working in betting shops, and running bingo halls—valuable contributions to the economy indeed.

The planners panicked when they saw a few million tons of coal going on the ground and produced the wrong solutions. To remove men from the production of wealth to wasteful servicing or the manufacture of non-essential consumer goods was always the wrong answer.

In the middle of 1966 the industry seemed to be at the centre of a process of endless contraction. Introducing the Coal Board's Annual Report for the year ended that March I looked ahead for a period of five years and forecast that in that time the reorganisation of the industry would be complete. I prophesied that pit closures would be finished and that the industry would be entering upon a period of stability.

All this has now come about. In fact it was achieved several months ahead of schedule. There were only one or two pit closures in 1971; all were due to exhaustion of reserves. Confidence was returning, as evidenced in the most convincing way. More people were joining the industry and fewer were leaving it, with the result that for the first time in thirteen years total manpower rose steadily. Productivity, which had been climbing at a rate that was brisk by average standards for British industry but sluggish for us, once more started to rise. Coal was able to compete on price with most of its competitors and beat the others. But back in 1966, when we had just declared a deficit on the year of nearly £25 m., when we were apprehensive about the fuel policy that the Ministry of Power had been working on for so long, and when labour wastage was at the highest rate since nationalisation, it took considerable confidence in the rightness of our policies to make such a bold prophecy. But the essential need was to rebuild confidence, and looking back it seems to me that I was always in the position of having to do this and to correct the excesses of the Ministers and their advisers. They quite often seemed to forget the need to give confidence to the people of the industry.

I had adopted an output of 200 m. tons a year as the industry's objective even before I went to the Coal Board. Many outsiders were arguing that if we got rid of our 'marginal' pits, the result would inevitably be a more prosperous mining industry. On many occasions I demonstrated that this just was not so, that it was vitally important to maintain as big an industry as we could. I spelt this argument out when I gave the Cadman Memorial Lecture to the Royal Society of Arts in January 1963. I pointed out that the industry paid about £43 m. a year in standing interest charges before it could even begin to talk about net profits. There were also about £185 m. a year of other fixed overheads that could be spread comfortably over 200 m. tons a year, but if we fell below that figure, our costs per ton would inevitably rise. On an output of 200 m. tons, standing overhead charges were roughly 22s. 6d. a ton, but on 180 m. tons they would rise to 25s. od., and on 150 m. tons to 30s. od.

But that was not even half the story. The greater part of our labour—all workers in fact, except faceworkers—constituted a cost which did not vary much with output. So if the industry lost 10 or 20 m. tons of its market, and therefore 5 or 10 per cent of its income, it would not reduce its costs in anything like the same proportion. Conversely, an expansion in our market of 10 or 20 m. tons within our existing capacity would reduce our costs appreciably by giving us that marginal tonnage as a bonus. I made the point, of course, that there were also important arguments on the grounds of humanity why output should not be cut rapidly.

I also showed that the shutting of a pit meant virtually the loss of its reserves for all time. In a closed colliery the strata converge, water seeps in, and in a few weeks nature can undo years and indeed decades of human work. This was a factor that was not always appreciated: people tended to think that a pit was like a factory and that you could close the front gate, put a lock on it, and go back and re-open it again years later when it might be needed. There can be no going back into the pits closed unnecessarily. Any reserves that they contained are completely lost.

It was in that lecture in 1963, incidentally, that I made an early public reference to the activities of the now redoubtable Organisation of Petroleum Exporting Countries. It was fashionable, in ministerial and official circles in those days, to be somewhat patronising of OPEC, but the gentlemen who held those views have now come to see the wisdom of my warning eight years ago. I stressed the inevitability of increased oil prices arising from the changing trends of supply and demand and from the need to finance future exploration.

Some of the figures for oil reserves I gave in the paper were criticised by a member of the audience. This was John Davies, then of Shell Mex and BP Ltd, later Director-General of the Confederation of British Industries and currently the Secretary of State for Trade and Industry (the Minister responsible for the nationalised industries). John Davies was the tenth Minister to whom I was answerable during my ten years with the Coal Board. Another warning given during this lecture—and this was one of the things that irritated John Davies and other oil industry representatives—was that the oil industry would quote very attractive prices to get into markets that it coveted. Once it had the business firmly sewn up, however, there would be nothing to stop it exploiting its position by increasing prices rapidly, unless there was a healthy coal industry for it to contend with. This process had already happened in West Germany at the time of this lecture. I showed that because they were active on an international basis, the oil companies could recoup temporary losses in one place by profits in another part of Europe. This was illustrated with a quotation from a West German newspaper, *Rheinische Post*, which a few days before had said: 'Imagine house coal prices rising within six months by 30 per cent. What a row there would be; what catastrophic effects on the standard of living would be asserted. In fact, the domestic consumer's fuel for his oil-fired central heating today costs 30 per cent more than it did six months ago. Since June 1962 the fuel oil prices had been rising steadily. . . . Such price increases are without precedent. . . . The time when domestic fuel oil was cheaper than domestic coal appears to have gone.'

Similar stories could easily be written now. If we substituted twelve months in the first sentence of that quotation for six months and 66 per cent for 30 per cent, we would have an accurate description of what happened here in 1970/71.

Although we had expected that it would take at least two years longer, our 200 m. tons a year target was reached in 1963. This early success was mainly attributable to the industry's achievement in raising the proportion of the output got by power-loading (that is, fully mechanised) methods. We had proved that our output objective, thought by many people to be quite impossible, could be attained. The fact that we were later unable to maintain a market of this size was the main reason for the industry's inability to win an even longer period of stable prices than we did.

I have always believed that it does people's confidence a world of good to know that they are working for a profitable organisation. This argument is just as valid in a nationalised industry as in private enterprise, although there are many Labour MPs and some others who try to argue that state enterprises like coal are a service to the community and profit is a secondary consideration. They should try that one on the Treasury, even under a Labour administration.

Because of this conviction that good financial results had a good effect on morale and confidence, I repeatedly declared my aim to break even in my second year in the Chair—1962. One of my former political colleagues accused me of being obsessed with this and I sharply criticised him. After all, the Coal Industry Nationalisation Act, which he and I had both supported in Parliament, required the industry to break even over a period of good and bad years. But, of course, I wanted to do much more than just show good financial results. As a supporter of the nationalisation of the mining industry it was necessary to show that coal could be a model of efficient public enterprise. I agreed with Herbert Morrison when he said that it was vital that national ownership should show itself superior to private enterprise in all-round efficiency.

Obviously we couldn't hope to pull back into profitable

working in 1961, but we did manage to slash the deficit after interest by £6 m.

Naturally, having stuck my neck out and aimed for a break-even result in 1962, I was impatient to see the figures for the first half-year. In fact, they couldn't have been better. After paying nearly £22 m. in interest charges to the Ministry of Power, we still had a surplus of nearly £9·5 m., compared with a surplus of £2·6 m. in the first half of the year before. This result did an awful lot for the confidence of the industry.

At that time a tribute I especially appreciated came from John Freeman, a former colleague in the Parliamentary Labour Party, and later appointed by Harold Wilson to be High Commissioner in India and subsequently our Ambassador in the United States. In the *News of the World*, Freeman wrote: 'Alf Robens has never forgotten that he is one of the workers. As a trade union official he used to represent the shop assistants who suffered —they still do sometimes—from grievous exploitation by the unimaginative little tyrants of retail business. Alf really feels how human beings can suffer from boardroom decisions. And his first care at the Coal Board has been to take the miners into his confidence. They haven't always agreed and sometimes they have resisted his plans but they've never doubted his goodwill. . . .'

When I gave the annual New Year Press Conference I was able to announce that when the accounts for 1962 were presented officially later in the year they would show that we had a small surplus. We had done the trick. I knew that many people in the industry had not believed that it would be possible. Fortified by the excellent results I was able to talk challengingly about the competition from oil and I warned that it would be a mistake to assume that it was always going to be cheap.

The value of productivity increases was vividly shown in 1962: output per manshift went up by nearly 8 per cent and this made it possible for 20,000 fewer men to produce 8 m. tons more coal than in the previous year—and from 450 fewer coalfaces.

The industry had recorded its first surplus for six years, but this did not disclose the full extent of the financial improvement that there had been. We had been able to set aside a good sum to reserve so that we would have a fine start in our aim of keeping prices stable at a time when almost all costs were rising.

A Government White Paper the previous June had imposed

an obligation on the Coal Board to set aside £10 m. a year for what was called 'fixed assets replacement'. This sum was intended to cover the higher prices that would have to be paid to replace assets that we used up. In other words, it was an allowance for inflation.

Knowing what I did about the provisions we had made, I had no hesitation in saying that we would, in 1963, be able to meet this obligation, to maintain price stability, and even make some price reductions. But I again made it abundantly clear that all this would depend on our success in producing and selling 200 m. tons of coal a year.

In fact, the industry had vintage results for the second successive year in 1963. Our report covered a fifteen-month period, as we had changed from a calendar year to a year that finished at the end of March—a more usual accounting arrangement.

Productivity rose by 7 per cent—again far more than the national average. Though our costs had risen sharply, we had covered them because of the better productivity and we again had a small surplus after meeting all our interest and depreciation charges. Exports, at a time when overseas business was vitally important, were up by about two-thirds. And we had been able to pay better wages. Again, most of these achievements had been possible because we had been able to keep our output and sales at around the vital 200 m. tons.

Presenting that Annual Report to the Press, I again warned that coal prices would have to go up unless we could maintain the industry at about this size. While the retail price index had gone up by 13 per cent during the last four years, we had managed to avoid a general price increase.

In six of my ten years we had a surplus after meeting all interest charges. In eight of the ten, we had an operating profit at the pits and in ten out of the ten we showed a profit on non-mining activities. Incidentally, this last item has shown steady and satisfactory growth every year since 1965/66 and the profit from these activities rose from £9·3 m. to £20·6 m. in 1969/70.

Over the decade the industry as a whole earned profits totalling £290 m., and had to meet interest charges of £342 m. This meant that the net result was a deficit after interest of £52 m. I suppose that sounds a lot of money, but in fact it represents 0·7 per cent of total turnover. If we had wanted to, we could have covered this by, for example, chopping off the most expensive 1·5 m. tons of output. It could have been achieved by shutting more pits, or by an additional increase of 4d. per ton.

It was in 1968/69 that our financial results, despite a record increase of 9 per cent in productivity and an increase in coal consumption, started to deteriorate. After paying interest charges, we had an overall deficit of almost £9 m. Inflation had begun to bite hard and it was quite impossible to get the increased productivity to provide for its mad gallop, since over 50 per cent of the cost of a ton of coal is for wages and related charges, despite the advance in mechanisation.

In some of our profitable years it was a near thing that we managed to squeeze into the black. Several times we had results for the first half-year that seemed to make a deficit on the full year inevitable. However, we used the half-year's figures to show everybody in the industry how the losses could be recovered. And, hopeless though it seemed in some of the years, we managed it by the skin of our teeth.

Never was there a more dramatic financial recovery than in the last of my ten years. At the half-year, we seemed certain to finish with a deficit of at least £10 m., and possibly even £15 m. It looked as though we were moving to a disappointing financial finish.

However, productivity recovered well after the strike the previous autumn and for the first time ever exceeded 47 cwt. per manshift for every man employed in the very last week of the financial year. Attendance had improved and our finances had benefited from the huge rundown in our undistributed stocks. Still, it hardly seemed possible that we could have pulled back the earlier losses.

Imagine my pleasure then, when David Clement, the Finance Member of the Board, gave me a first sight of the figures which showed that we had managed a tiny surplus after all.

Although the Board closes its books at the end of March each year, the financial results are not usually disclosed until the Annual Report and Accounts are published by HM Stationery Office about the end of August or beginning of September.

This was one year when I felt it impossible to wait that long. I was due to leave at the end of June. The results were given to our May Board meeting and I made an announcement to the Press immediately afterwards. I wanted the whole industry to know we had achieved what we had been working for and to share my satisfaction. The news would increase the feelings of confidence that were slowly returning, as the improved recruitment and reduced wastage figures were already showing. And I needed some personal cheering up at that stage. This time it was a morale booster for me, too.

Between 1960 and 1970 we increased productivity (as expressed by output per manshift) by 58 per cent. This average increase of 6 per cent a year far exceeds the increase in the rest of British industry as a whole. Indeed, had our record been matched throughout the economy, the nation's condition would have been enormously better than it was. There would have been no anxieties about balance of payments, the increase in real wealth would have made it possible for us to have far more houses, schools, hospitals and other desirable improvements. Income tax could have been reduced long ago and by a substantial amount.

How was it done in our case? The increase in mechanisation that I have already mentioned was the big factor. But it was not just a question of putting the machines in and then sitting back and waiting for the results to flow. Different kinds of machines had to be designed for different conditions. Their introduction had to be properly planned and prepared. And above all it was necessary to train people in their use. Then, when all this had been done, it was necessary to make sure that they were intelligently used.

In the days before mechanisation, it was usual for the machine to make one cut of the face on each shift and for the men to fill that off with shovels. When the power-loading machines came in, there was a tendency for the men to say: 'Well, we used to give you one cut. Now we've got the machines we'll give you two.' But in fact the machines were capable of doing not two, but four or more strips per shift. It took some time to convince everybody of this. They thought that because twice as much coal was coming off the face as before, the machines must be paying for themselves. That men eventually came to agree with us, despite the fact that being on day wages they wouldn't get any more money for doing a larger number of strips, was a great tribute to their spirit. There are not many industries in the country where productivity has gone up so rapidly, and yet this was all achieved at a time when we eliminated the hateful carrot-and-stick methods of payments that are traditional with piece-work. People have argued that because we have closed so many pits, our increase in productivity was inevitable since the pits that would go out would be the ones with the worst performance. In fact, we calculated that the rise in productivity attributable to this factor was a good bit less than 20 per cent of the whole increase.

In 1960, only 37·5 per cent of the total output was power loaded—that is, it was both cut and loaded on to conveyors by machines. Nevertheless, by the end of March 1964, almost three-quarters of the coal produced was mechanically cut and loaded. In the succeeding years progress, not surprisingly, was less rapid. The easiest and best faces were mechanised first; the less easy ones took rather longer.

By 1970/71, virtually every face that could be mechanised had been dealt with and 93 per cent of all output was power loaded. At the beginning of the decade conventional working meant machines to cut the coal and men to fill it off with shovels. By the end, what had been regarded as advanced at the beginning had become the new conventional method. Not only did the machines cut and load the coal, but, as I mentioned in Chapter Five, on many faces we had roof supports that advanced themselves at the touch of a switch and conveyors which were snaked over mechanically as the machine went past on its journey along the face.

Mining today is certainly no soft job. For the men simply to travel along the face four times or so during a shift attending the machine is still physically hard going. Nonetheless, much of the hard slogging has been taken over by the machines. Work at the coalface is no longer the physically exhausting job that it used to be. The pick and the shovel have become almost museum pieces on most coalfaces. Coal-mining is certainly one industry where the machines have been welcomed. As I often said: 'There are no Luddites in coal-mining.'

In the days when the coal was cut by machines but filled on to conveyors by hand after being blasted by explosives, it was customary to produce only on one shift out of the three. Nevertheless, a great many men were at work on the other two shifts moving forward the face supports and equipment, and advancing the roadways.

The introduction of machines that both cut and loaded the coal meant that less time was needed for preparation. So we were able to produce coal on two and three shifts from the same face instead of only one. By the end of my ten years' spell, between 80 and 90 per cent of all longwall faces were working on two or more shifts, compared with only about 60 per cent at the beginning of my Chairmanship.

One of the tremendous, unqualified achievements, and one that will certainly endure, was the acceptance of the idea that piece-work should be abolished. Ever since the industry started, the men have been accustomed, when they were working at the face or on other jobs where their effort could be measured, to being paid by results.

Now men are quite prepared to work on coalfaces that produce 3,000 tons of coal a day for the same wages that they got when that face produced only 300 tons a day.

Technical progress is one thing. It is quite another thing to get men to change their attitudes completely like this. I know many men whose judgment I respect and who have been in the industry all their working lives who said that this change would never come about—at any rate, not without sacrificing productivity. In fact, some of the biggest productivity gains between one year and the next have been made since we got rid of piece-work on the power-loaded faces. And the other big gain, of course, has been that all the arguments that used to go on at collieries about allowances for this, that and the other, about the rate that was to be paid for new jobs, about whether a man had been prevented from achieving a reasonable wage through circumstances beyond his own control—all these arguments, that used to sap the energy and swallow the time of managers and underofficials, union officials and men, have been swept away. We have enormously reduced the amount of coal lost through disputes of this sort at a time when virtually the whole of the remainder of British industry has been suffering an increase. This great advance had, of course, to be worked for. And it took a long time. But the results have over and over again repaid the trouble that was taken.

During the years before piece-work was abolished on most

coalfaces the average annual loss of output through disputes was 1·4 m. tons. In money terms that would be about £7 m. down the drain every year. By far the greatest source of disputes were the piece-work arguments and the arguments that arose from them about allowances and so on.

The improvement after the National Power Loading Agreement was quite spectacular. The average loss of 1·4 m. tons in the preceding eight years was cut first to 439,000 tons, and then in the next year (1969) to only 328,000 tons. That the figures soared in 1970 to 2·8 m. tons and in 1971 to more than 3 m. tons, was not so much due to anything that went wrong with our own wage arrangements. The unofficial strikes in the autumn of both 1969 and 1970, which I have written about in Chapter Two, were a reflection of the industrial unrest in the country as a whole. We were unable to isolate the industry from forces outside it. But nothing can take away the tremendous achievement that had been marked up in this field of industrial relations. In the last weeks of my term at the Coal Board we concluded negotiations with the National Union of Mineworkers which brought another 50-odd thousand people, working mainly in non-mechanised places, on to day wages. The industry had seen that the process could be carried through while still keeping productivity climbing. And another of my personal ambitions had been achieved.

For the first seven or eight years of my ten, we were able to follow what was in fact a wages policy of the kind sought unsuccessfully by Governments of both parties. Pay increases were covered by our increased productivity. In fact, in some years they were more than covered. Productivity increases also meant that for most of the time we were able to absorb increases in the price of materials and services that we had to buy ourselves.

Almost all our settlements gave more money to the men on the minimum wage than the better paid got. We were also able to get rid of many anomalies, particularly by the abolition of piece-work, and to make a start on eliminating the differences between the most highly paid coalfields and the less well-paid. This policy obviously depended on the strong leadership of Sid Ford and Will Paynter, the national officials of the Mineworkers' Union, who achieved miracles in persuading some people to accept small increases whilst their less fortunate colleagues in other areas caught up with them.

All this meant that we were able to achieve, in the early years anyway, a very high degree of price stability. So the customers also shared in the benefits brought about by increased mechanisation.

Although we managed to show a small surplus again on the financial year 1964/65, I had to give warning, announcing the results, that we should shortly have to have price increases. This was the end of a remarkable period of price stability. Pit-head prices of most industrial coals had remained unchanged for five years, and of domestic coals for three and a half years. In the rest of the economy, prices had been rising steadily. For example between September 1960 and December 1965 the wholesale price index went up by 15·2 per cent.

Hardly anyone gave the Coal Board any credit for this remarkable achievement. I've always found people ready to shout about price increases, but not so eager to give credit for price stability.

We managed to achieve a reduction of 2½ per cent in the costs per ton in 1967/68; this was only about the second reduction between one year and the next recorded since nationalisation. There was also a reduction of about 1 per cent in the average price we charged per ton. What happened was that we were producing less of the higher-priced coals and a bigger proportion of our output was going to the lower-priced markets. The improvement in production costs that year was another clear bit of evidence about the value of higher productivity. During the year as a whole output per manshift increased by 6·6 per cent, but during the second half of the year the rise was 8·3 per cent.

However, as I have said, it is not possible to isolate one industry from the rest of the economy. During the last couple of years, as the NUM leaders claimed during the wage negotiations in the autumn of 1970, the miners certainly lost ground relatively by comparison with people in other industries. Because just over a half of our total costs are wages costs, we were very badly hit by the wage inflation of 1969 and 1970. All this coincided with a temporary failure of productivity in the pits to rise as rapidly as it had been doing, and as rapidly as we had hoped.

The result was the succession of price increases which came

during my last twelve months. The unofficial strikes of 1969 and 1970 also added grievously to the industry's financial burdens. But they too were attributable to the new militancy and the galloping wage inflation that was going on all round us. For far too long too much of British industry has been buying industrial peace through the cheque book instead of insisting that the increased productivity must be there to pay for increased wages. I know these are very obvious statements. But that does not stop them from also being very true.

Price increases were anathema to all of us; it made it more difficult for our marketeers to sell coal and thus harder for us to maintain the size of the industry. Price increases ultimately meant more pit closures. On the whole our record on price increases was not unreasonable. Domestic coal went up by 4·5 per cent a year compared with an average annual rise of 4·7 per cent in the retail price index. The power stations had increases in their coal prices exactly in line with the average increase in the wholesale price index. The average price increase for all our coals has been much less than in the average price indices because of the switch to lower-priced markets.

Another important financial achievement was the way we managed our modernisation and re-equipment programme. During the ten years we invested £860 m. in the collieries and ancillary activities, and provided £710 m. of that out of our own resources. We did not make heavy claims on the nation's capital resources.

But, as I said earlier, our report on progress is concerned with much more than the financial results and the prices we charged. In 1960 there were 602,000 men on colliery books. By the beginning of 1971 the figure was down to 285,000. In a single year (1968), 25,000 men were made redundant. The statistics themselves are staggering enough when one begins to try to assess the effect on people. It is almost too much for the imagination. And again, the rundown in people as well as in the number of collieries has proved to have been unnecessarily severe. When the demand became stronger again, we were struggling to increase the labour force for the first time in more than a decade. Most of our recruits came from the mining areas, of course. Indeed, most of them are the sons or brothers of miners. To go from a rapid rundown and to try to reverse trends is very difficult indeed. Confidence in the future of the industry was savaged by successive Governments and especially by the Labour Government's Fuel Policy in November 1967. The Coal Board are still having to struggle with the aftermath of that devastating blow.

It is one of the industry's finest achievements, however, that this enormous reduction in the number of people employed was accompanied by the minimum social hardship. Government provisions helped to cushion the worst economic effects of redundancy. But the policy of bringing new industries to the mining areas worked with pitiful slowness and at great expense to the community. As I write, more than 900,000 people are unemployed in this country. The rate is especially high in some mining districts. The tragedy is that a great many men now drawing unemployment benefits could have still been at work producing coal. It angers me when I think that many of those unemployed miners are the victims of a nuclear power programme that was much bigger than was justified. They lost their jobs, not because the coal they produced was uneconomic by comparison with nuclear power, but because we were thought to need a white-hot technological revolution.

The progress recorded in this account of ten years in coal was the result of a lot of hard work by a lot of dedicated people. And despite the difficult and dangerous character of the industry, it does inspire dedication in the people who work in it. I have been involved in public service for virtually the whole of my working life, and have been blessed with energy, determination and a certain amount of skill in influencing and inspiring other people. I believe I have used those talents for what they were worth to lead an industry to which I was emotionally bound. I have certainly enjoyed great satisfaction in working with as fine a crowd of people as are to be found anywhere in the world. It was a great team job and I was very privileged to lead it. The industry succeeded in spite of the experts, civil servants and Governments. And the whole country benefited.

In a profile which he did for the *Daily Mirror*, Matthew Coady wrote: 'Miners' leaders, whatever their criticisms of Robens, would agree on the value of his work. Any other single individual, given the job of running down the industry, would have been faced with violent upheaval. Under his leadership this has not happened.

'The fight on this perhaps has been totally ruthless. His knowledge of Whitehall has taught him every trick in a sophisticated book. Civil servants have been harried, ministers lectured and Press leaks deftly engineered to serve coal's cause.'

I admit the ruthlessness in that context of fighting for the industry. It has lost me some old political friends. But I reckon it has gained me something I value enormously, and that is the respect, gratitude and, yes, affection of the vast majority of the people in the industry.

16

The Decision to Go

Mining is much more than an extractive industry, bringing mineral out of the ground. Each colliery has its own tightly knit community, and the old and the young, the sick and disabled are all part of pit life. The social life of the village and all the good voluntary work that is done by the men and their families, and by those who serve them on local authorities and other bodies, are strongly influenced by the pit. It is quite impossible, even if it were desirable, to divorce oneself as Chairman of the Board or Manager of a pit from the emotional and sentimental attachment which close working with miners produces.

When my first five-year term was over, therefore, I needed no persuading to do a second. A third spell would have completed my working life, and when I was reappointed in 1965 I had privately decided to do a total stint of fifteen years if I were invited to do so.

I was not short of extremely tempting job offers from elsewhere, but I was not, and indeed never have been, in pursuit of money. All my life I have thoroughly enjoyed the work in which I have been engaged. In mining this was even more true.

Only politics could have tempted me away and by that time it was clear that I was unlikely to receive any invitation from the occupant of Number 10.

So I was really all set to go on as long as I was wanted. There was so much to be done in the years that lay ahead. The pattern of world energy was changing and the newer technological developments in engineering seemed likely to be capable of application to the old industry of coal mining.

The second five years seemed hardly to have been entered upon, when I was past the half-way mark. And soon 1970 dawned, the last year of my second term.

The autumn of 1969 had been difficult for a number of reasons but mainly because of the outbreak of unofficial strikes. One was glad of the end of the strikes and the short break over Christmas and the New Year. But 1970 was far from being an easy year, and as the winter sped by my mind inevitably turned to the future and the possibility of a third term.

The change of Government came in June 1970. Harold Wilson had misjudged the timing of the election; he might well have succeeded in an October election. Certainly he had seriously misjudged the temper of the people. He ran the election as though it were a Presidential campaign and he failed miserably, leading the Labour Party to defeat.

The new Government soon got to work and built up a new Department of Trade and Industry, which swallowed the old Ministry of Power whole. It was made clear that they were going to follow a policy of disengagement from industry and to prepare plans for the 'hiving off' of part of the nationalised industries.

The Coal Board, whilst it had been gradually reducing its mining business under the Labour Government, had been expanding in other directions by diversifying into related commercial activities. By this time the ancillaries were contributing about £13 m. a year profit to the kitty; without them, coal prices would have had to be increased in total by that amount, and raising coal prices was not likely to improve the market prospects.

There were a number of different reasons for our ancillary activities. For instance, one of our very successful ventures came about because of the grip that private enterprise had upon manufacturers, preventing those manufacturers from trading with us under the threat of a boycott.

The way it happened is quite an object lesson. In our drive to keep the domestic solid fuel market against competitors like gas and electricity, who had sales outlets in every town, we decided to open similar showrooms where we could display and demonstrate the latest in solid fuel appliances that complied with the Clean Air Act. These showrooms were not inexpensive but, besides helping us in our marketing operation, they were of great service to the public, enlarging their consumer choice. We sought to sell the appliances from the showrooms, but the manufacturers were not permitted by the distributive trade to supply the Board and threatened to refuse to buy not only solid fuel appliances, but all the other things like gutters and downpipes, sanitary ware and so on manufactured by the same companies. The consequence was that our salerooms were only activated at half rate—we could only show prospective customers the appliances and recommend them to a list of hardware stores or builders' merchants who might or might not supply.

This was not only an expensive way of selling, it was also ineffective. We could not follow the customer through as gas and electricity could and in fact were left wringing our hands, as he or she left our showrooms discouraged and lost to us for ever.

I came to the conclusion that the only way to circumvent this threatened boycott was to acquire a builders' merchants' business ourselves, and this we did. When I travelled to Rome on one occasion, my fellow passenger was a merchant banker, to whom I explained the situation and invited him to let us know if he heard of any suitable prospect.

As luck would have it, the opportunity came quite quickly and we became partners, through British Sisalkraft Ltd, the subsidiary of an American company, St Regis Paper Company, in J. H. Sankey and Son Ltd, an old-established builders' merchants they had acquired with some other business and didn't particularly want. Today the Coal Board own 77 per cent of the equity and British Sisalkraft the remainder. It was a a small business to begin with, but it was apparent that a large number of builders' merchants were only too anxious to become partners in a larger group. As the spread of shops increased we no longer needed our own showrooms, so where it was a commercial proposition we merged the showrooms with Sankey and opened no more. Today Sankey is about the third largest builders' merchants in the country, profitable and providing the Coal Board with a nation-wide chain of solid fuel appliance showrooms where the customer can get the full service covering purchase, installation and easy terms, under the one roof.

All this came about not because of a giant publicly-owned enterprise behaving badly, but because of threats from the trade.

Another very highly profitable venture was our entry into the North Sea gas operation. The Coal Board were not, of course, newcomers to drilling in the North Sea. The Board were in fact the first ever to do this. As far back as 1955 the Board's drilling rig was in action in the Firth of Forth. Three years later we put down the first borehole in the North Sea and subsequently operated up to six miles out off the Durham coast. Although we were prospecting for coal under the North Sea and not drilling for gas we had what everyone else lacked, the experience of actually drilling in those difficult waters. We were also able to offer the largest team of specialists, headed by George Armstrong, our very able Chief Geologist.

The first licenses for drilling in the North Sea were granted in September 1964, and it was not long before we became involved. At a New Year's lunch I happened to meet Dr Jerry McAfee, President and Chief Executive Officer of Gulf Oil. I told him of our own interest in the North Sea, doing the geological exploration for our coastal collieries, and I suggested to him that of all possible British partners not already in oil we were the only ones with anything to offer. He showed much interest. By April we were discussing draft heads of agreement for a partnership, and took formal steps to further the business.

Our major difficulty was that the NCB had no statutory powers to drill for gas, so I told the Minister, Fred Lee, that Gulf were interested in some kind of partnership with us and we ought to be given statutory authority to enter it. He agreed that this was worth doing, and that the nationalised industries should become more involved in Continental Shelf exploration. However, at that time he was unable to give us any specific assurance that we would be given the necessary statutory powers. So we were not in a position actively to look for partners, but this did not matter as we would not go ahead unless the Government thought it desirable.

We continued our negotiations with Gulf. After the Minister announced that a second batch of concessions would be licensed with a clear understanding that consortia containing British interests would be welcomed, we also started negotiations with Allied Chemical, another American company, who approached us through an intermediary. Both sets of negotiations led to our being given an option to join in partnerships with the companies concerned, so long as we received the necessary statutory consent and were thus able to take up our options in time to meet the companies' drilling programmes.

There followed an anxious period of waiting, as Parliamentary time had to be found for the enabling Bill. Eventually it became clear that we would not secure this authorisation until the very last minute, but get it we did.

In September 1966, I was in the States and called on L. F.

McCollom, then Chairman and Chief Executive of the very successful Continental Oil Company. The end result of a number of conversations was that we became partners in what has been one of the most successful groups in the North Sea. Our rich Viking Field will supply about 15 per cent of the total gas market and during the twenty-five years or so of its life will bring into the Coal Board's coffers about £200 m. So the financial return will make my salary for the whole ten years look like petty cash.

In a modest sort of way we had always been in chemicals. The big coke ovens produce as a by-product crude benzole, which for many years was mainly used as motor spirit. With the abolition of subsidy on home-produced petrol, it was no longer economic to go on making it. We had to do something with our benzole but certainly had no desire to sell it at a give-away price. That meant that we had to take a further step into the chemical industry. We did so in association with one of the big steel companies who were faced with a similar problem. Together we built a modern benzole distillation plant which neither would have been justified in doing on their own.

We also partnered with another nationalised industry, but not a British one. The Dutch State Mines have done extremely well out of chemicals from coal and have earned vast profits from this source at a time when coal output from their collieries has been shrinking. Indeed, they now operate only one mine.

Our jointly owned plant makes Caprolactam, which is the raw material for Nylon 6. The feedstock for this process is cyclohexane, which is made from benzene, derived in its turn from crude benzole supplied by the Coal Board and the British Steel Corporation from their coal carbonisation plants.

We shall certainly do well financially out of the chemical plants we have built, but it is a commonly held misbelief that chemicals from coal can be the salvation of the industry. It is possible to make an awful lot of chemicals from a comparatively small quantity of coal. I suppose if you were able to develop overnight all the plants needed to make all the possible chemicals from coal that could be sold economically, you would not have a total market for more than one or two million tons of coal a year. Leslie Grainger, the Board Member for Science, used to say: 'If this lady [meaning coal] is not for burning, we are not going to have much of a business.'

The Coal Board were among the first industries to develop the use of computers. With our huge and complicated pay-roll work, they were a boon to us. We soon had seven computer centres, one in each major coalfield, and, when they had got used to belting out all the industry's pay tickets, they still had plenty of capacity for other work. So we gave them more work to do. For many years now the computers have invoiced all our customers, kept all our stores records and so on, as well as paying the wages. And we have gone on loading fresh jobs on to them.

By about three years ago it was obvious that we would never, with Coal Board work, exhaust the potential of the machines we had to have to carry on our business. So we decided to sell the spare capacity to outside users under the sales title of NCB Computer Power.

This work has proved a profitable outlet and is still growing in volume. NCB Computer Power not only offers time to industrialists and others who have insufficient work to justify buying or renting a computer of their own, but we also train suitable people from those clients at our main centre in Cannock, Staffordshire.

The sale of our spare computer capacity was not only an attempt to utilise our resources at a profit. We also provided a very valuable service to the client. For example, for a few coppers (old-style) per employee we handled a company's weekly pay-roll work. Certainly, they save money by comparison with paying their own wages staff or by renting their own computer. And in many cases, the people we worked for were small firms whom we relieved completely of any need to employ staff of their own on pay-roll work or, come to that, stores accounting, stock control, invoicing and many other jobs that used to employ throughout the country armies of clerks.

An even more spectacular use of our computer skill has been International Reservations Limited, in which we are partnered by an American company already well established on the other side of the Atlantic. This enterprise is a service for booking hotel rooms. People who ring a nearby reservation centre can, within seconds, make the bookings they need. There are now 750 hotels in Great Britain and Ireland, between them providing 50,000 rooms, on our list, and, including those in North America and Switzerland, brought in by our partners, more than 7,000 hotels and 775,000 rooms.

This idea came from the report of the Economic Development Council for the Hotel and Catering Industry. I was a member of the National Economic Development Council at the time and immediately co-operated with Sir William Swallow, who presented the report as Chairman of the little Neddy.

The service has recently been extended to provide also for hire car reservations and has access to 8,000 cars and 70 locations. I am convinced that before many years are past no one will ever dream of making reservations in any other way than through a system of this kind.

I suppose that on the surface this does seem to have been a surprising business for the coal industry to involve itself in. But we are, after all, a public undertaking. We needed to acquire the computers for our main business. Why should not the public get a return from having the computers fully used? Our gross annual income from our bureau activities now more than covers the cost of all our hardware. Furthermore, the income from this and our other ancillary activities helps to keep down the price of coal, as I have said. Which helps us to supply the country with more indigenous energy and keep more British miners in work.

Our biggest development into retail fuel distribution was in partnership with Amalgamated Anthracite Holdings Limited. This was formally established during 1966 under the name of the British Fuel Company. Later BFC formed a further partnership in the West of England with Renwick, Wilton and Dobson (Partner) Limited. The solid fuel trade was further strengthened in another part of the country by our formation of the Lancashire Fuel Company Limited.

Encouragement for us to extend our activities further in the field of coal distribution had been provided by a report of the National Board for Prices and Incomes published in September 1966. It called for amalgamations into larger distribution organisations, a rationalisation of ordering, collection and delivery, and an extension of NCB retailing if the industry failed to re-organise itself on the lines recommended. So in extending our distribution network and by entering into partnerships with firms of distributors, we were carrying out a clear recommendation made in the national interest by an authoritative and impartial body. Naturally, as producers we leaned over backwards to avoid giving special treatment in supplies to our own retail service or to our partners, as against the independent distributors.

All our ancillary undertakings were associated with our main business. We had not become a conglomerate or moved into completely unrelated fields. What we did was to utilise our own resources, physical and human, our own particular expertise, which a shrinking mining industry had made available. The aim was to produce profits which we could use to help stabilise the price of coal and thereby improve the financial position of our primary business.

Obviously Labour Governments were quite keen to see the nationalised industries branching out. The Government White Paper, *The Finances of the Coal Industry*, published in November 1965, gave the go-ahead for the Board to diversify by allowing us to invest about £75 m. (at 1964 prices) in the five years up to March 1971 on non-mining activities.

One of the first journalists to write about these activities of ours was Christopher Tugendhat, then a member of the staff of the *Financial Times*. Tugendhat said that the wider the range of uses developed for coal, the more chance there would be of maintaining its production at a reasonable level. And he added:* 'Faced with a similar situation most private enterprise companies would follow the same kind of policies as those of the NCB. But there is one big difference between public and private sector organisation causing great growing concern. In the private sector companies can only diversify if their ideas are financially sound, whereas nationalised industries can often get hold of the tax-payers' money on easier terms. If the NCB can justify its projects on financial grounds it should be able to counter the arguments against back-door Nationalisation, and recover some of its lost prestige.'

Tugendhat is now a Conservative Member of Parliament, and his party in office quickly sought to investigate our ancillary activities with a view to instructing us to hive off those that they thought we should no longer continue to be involved in, either as wholly owned enterprises or in partnership with other concerns.

And this was the beginning of the end for me. In the month following the General Election, on 30 July 1970 to be precise, I had a conversation with Sir John Eden, who was the Minister

* The *Financial Times*, 7 July 1966.

of State at the Ministry of Technology, later included in the Department of Trade and Industry, particularly responsible for the nationalised industries formerly covered by the old Ministry of Power. At the end of our business conversation, I raised with him the need for making a decision quite soon about the Chairmanship. I made it clear that I was not canvassing for the job, but that as my term of office was up at the end of the year or thereabouts, it was necessary for a decision to be made, first because the uncertainty was beginning to unsettle the industry (newspapers were starting to ask about my personal future), and if it were decided to make a change—a perfectly proper decision to make—then obviously I would myself require time to arrange my own affairs as I did not propose to go into retirement at sixty. He said that he fully understood the position. He would require to consult colleagues but August was a difficult month and he would have a further discussion with me as soon as possible. I also said that any new contract would require a salary equal to that of the highest-paid Chairman of a nationalised industry, as I could not accept a lower ceiling for my management personnel than the steel industry had.

It was a pleasant conversation and I left feeling that I should learn something by the end of August, or at the latest the first weeks of September. But the months went by—and with some considerable embarrassment. The pressure upon me by the Press, television and radio interviewers grew. I could give no answer that satisfied them or anybody else. The big wage negotiation boomed up and soon I was heavily involved in the detailed discussions, which were long and extremely difficult. The settlement was followed by unofficial strikes and the atmosphere was tense. Some writers then suggested that my firm stand on the wages front would be rewarded by a third term. Others took this theme up and I just had to stand by and keep my trap shut. In any case there was really nothing I could tell anyone. Nothing had been said to me since the end of July.

But other things had been happening. The industry had to have a new Act of Parliament. The 1967 Act had to be replaced by a new measure, since its provisions were due to expire at the end of March 1971. These provisions were very important because among other things they made special financial arrangements to deal with the redundancy in the mining industry.

Without a new Act we could no longer go on paying men benefits which the 1967 Act had authorised.

It is customary for all Governments to show the drafts of Bills such as this to the Chairman of the industry concerned, and to discuss them with him. This was no exception. The draft had eight clauses and after many discussions there was agreement on the provisions proposed. So far so good.

At the same time as discussions were taking place on the Bill, I was having talks with John Eden on the hiving-off proposals. My very strong and firm recollection is that we agreed to discuss what might be done about the ancillary undertakings and that we would bring in some outside specialist advice. In fact, a firm of chartered accountants and another of merchant bankers were named. I gathered that after careful consideration and further discussion we might come to some agreement as to how the ancillaries should be handled that would enable the Board to continue to have some trading benefit and at the same time accommodate the views of the Government about hiving off.

I was in fact a supporter of the view that we should go to the market instead of the Treasury for investment finance. Indeed I had urged this for the past five years but to no avail. I saw no problem in introducing equity capital and conforming fully to the Companies Act. This was not new either. As long ago as 1946, when I was Parliamentary Private Secretary to Alfred Barnes, then the Minister of Transport, he proposed the introduction of equity capital for the nationalisation of transport. His proposal did not find acceptance. I felt sure however that there were methods that could satisfy the Government's political requirement and our own business necessity, and I was content to agree to discuss various methods.

This is what I firmly understood was agreed. I was horrified and shocked to find, however, that the Coal Industry Bill, when published, dealt with the limitation of the Board's activities. Indeed it reduced them below the statutory authority of the original 1946 Act. I was thunderstruck. I had been a Minister when John Eden was in his political nappies, but had never experienced such peculiar behaviour.

The appalling feature of the Bill was the right for the Minister to issue a direction on all activities other than the mining of coal and its preparation for the market. Had the Bill become an Act, coal would no longer have been in business except by Ministerial decree. There would have been no right of appeal against any direction, no independent arbitration—not even Parliament would have been permitted to interfere. I could only think of two men in Europe who wielded in my life such dictatorial power in peacetime.

Ironically enough, when the Bill came before Parliament, it was that high Tory, Sir Gerald Nabarro, who rebelled against this extraordinary clause. He and Trevor Skeet, another Tory MP, voted with the Opposition and thus ensured that any direction objected to by the Board has to be approved by Parliament.

Thank God for the Nabarros of this world: he kept both his principles as a Tory, and as a democratic and a stout Parliamentarian.

But the Committee stage of the Bill was completed on 9 February 1971, and before that date was reached, an original draft had been turned into a very different Bill.

The strikes were over and work proceeding smoothly again when I was asked on 27 November to see Mr John Davies, Secretary of State for Trade and Industry, and Eden's boss. When I arrived I found Sir John Eden with him. After the usual pleasantries John Davies asked me if I would accept a third term. I replied quietly, but not without some inward emotion: 'I'm sorry but I cannot accept for two reasons. One is that I cannot accept the Bill and the other is that it is too late.'

I must draw a veil over the rest of our discussion, as the privacy of the conversation must be preserved. I can only say that my recollection is that nothing was missed out.

However I did agree to reconsider, and I did. The more I pondered, the more I realised how impossible it was for me to be the instrument for 'hiving-off'. In the world of builders there are two most important people. There is the architect whose task is to produce the plan to enable building to take place. Equally important is the demolition expert, whose job is to pull down and clear the site to enable building to begin.

Both important, both essential, but I am by nature an architect not a demolition expert.

And so it was time to go and on 5 January 1971 the following announcement was issued by Mr John Davies:

The Secretary of State for Trade and Industry (The Rt Hon. John Davies MP) announced today that, on 27 November, he invited Lord Robens, the Chairman of the National Coal Board, to continue in office for a further period of 5 years. Lord Robens informed the Secretary of State that after full consideration he did not wish to serve for a third term as Chairman of the Board, but that he would be prepared to remain as Chairman for a period of up to 6 months after the expiry of his present term while arrangements are made for the appointment of a successor.

I had willingly agreed to stay on whilst the new Board were appointed. Indeed I played an active part in the new arrangements. I was gratified to see the appointment of all the new full-time members from within the industry. In addition to Derek Ezra's appointment to succeed me as Chairman and Bill Sheppard's promotion to Deputy Chairman there were four new appointments: John Brass, formerly North-Western and Yorkshire Regional Chairman; Wilfrid Miron, formerly Regional Chairman for South Wales and the Midlands; Norman Siddall, formerly Director-General of Production; and Jack Wellings, Chairman of the 600 Company, as an additional part-time member.

So for me the end of an era, full of everything a man wants out of his business life. Success and disappointment have come hand in hand, but as a Lancastrian I understand that it's the warp and the woof of the cloth that make the pattern, and they cross one another at right angles.

This was as true of the comments on my departure as of those that greeted my arrival. There were a few sour observations from the expected quarters but they were a thin thread in the rich weave of hundreds of letters of good wishes from miners and their families, and dozens of warm farewell ceremonies like the one at Silverdale Colliery that I described earlier.

I am proud of my ten year stint and proud of the workmates who shared it with me. We have left our successors better prospects than we had to meet. And that's the test we all must face in life.

Statistical Summary 1960–1970/71

	1960	*1963/64*	*1967/68*	*1970/71*	*Change %* *1970/71 on* *1960*
Coal Production (m. tons)					
NCB Deep-mined	183·8	187·2	162·7	133.3	− 27·5
All Coal	193·6	195·2	170·9	142·4	− 26.5
Coal Disposals (m. tons)	200·0	197·9	164·4	150.7	− 24·7
Undistributed Stocks					
(m. tons) at end year	29·2	18·4	26·6	6·2	—
Number of producing					
collieries end year	698	576	376	292	− 58·2
Average Manpower ('000)	602·1	517·0	391·9	287·2	− 52·3
Output per man year					
(tons)	305·3	361·8	413·8	463·0	+ 51·7
Output per manshift					
(cwts)	28·0	33·4	39·0	44·2	+ 57·9
% of output power					
loaded	37·5	68·4	89·7	92·2	—
Disputes—Tonnage Lost					
(1,000 tons)	1,559	1,356	439	3,081*	+ 97·6
Accidents—Fatal and					
Serious					
Number	1,869	1,577	1,105	690	− 63·1
Rate per 100,000					
Manshifts	1·34	1·35	1·28	1·04	− 22·4

* Including a direct loss of 2,845,000 tons in the unofficial strike in October and November 1970.

Index